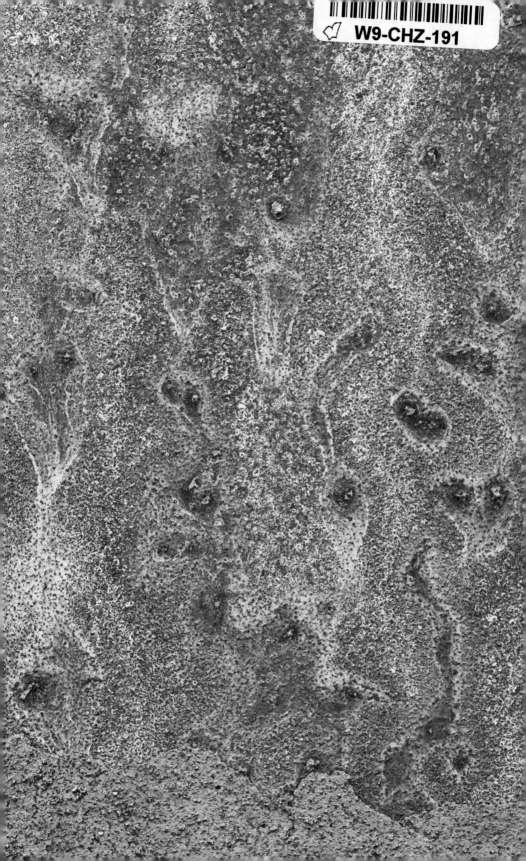

RUST
THE LONGEST WAR

JONATHAN WALDMAN

SIMON & SCHUSTER

NEW YORK   LONDON   TORONTO   SYDNEY   NEW DELHI

Simon & Schuster
1230 Avenue of the Americas
New York, NY 10020

First Simon & Schuster hardcover edition March 2015

SIMON & SCHUSTER and colophon are registered
trademarks of Simon & Schuster, Inc.

For information about special discounts for bulk purchases,
please contact Simon & Schuster Special Sales at
1-866-506-1949 or business@simonandschuster.com.

The Simon & Schuster Speakers Bureau can bring authors to your live event.
For more information or to book an event, contact the
Simon & Schuster Speakers Bureau at 1-866-248-3049 or
visit our website at www.simonspeakers.com.

Interior design by Ruth Lee-Mui
Map by Paul J. Pugliese
Endpaper images © Alyssha Eve Csük

Manufactured in the United States of America

10   9   8   7   6   5   4   3   2   1

Library of Congress Cataloging-in-Publication Data

Waldman, Jonathan.
  Rust : the longest war / Jonathan Waldman.
    pages cm
  1. Corrosion and anti-corrosives—History. 2. Corrosion and anti-corrosives—
Anecdotes. I. Title.
  TA418.74.W35 2015
  620.1'1223—dc23
  2014043291

ISBN 978-1-4516-9159-7
ISBN 978-1-4516-9161-0 (ebook)

For Mom and Dad,
and whoever bought that stupid sailboat

Only entropy comes easy.

—ANTON CHEKHOV

CONTENTS

PREFACE: A JANKY OLD BOAT

They say a lot of things about boats. They say a boat is a hole in the water that you throw money into. They say *boat* stands for "bring out another thousand." They say that the pleasures of owning and sailing a boat are comparable to standing, fully clothed, in a cold shower while tearing up twenty-dollar bills. Consequently, they say that the best day of a sailor's life, aside from the day he buys a boat, is the day he sells it.

Ignoring all of this wisdom, I bought a forty-foot sailboat. This was at the end of 2007. She was in San Carlos, Mexico, at a pretty marina on the Sea of Cortés. There were palm trees and haciendas, with deep sparkling water to the west, a rugged volcanic tower to the east, and an immaculate Sonoran sky overhead. With two friends, we split her three ways. I'd thought she was a bargain, but the marina was more *bonita* than our new boat.

Our sloop was thirty years old, and showed her age. There were little rust rings around every screw on the deck, rust stains on the stanchions, bow pulpit, and pushpit, streaks of rust down the topsides. A white powder surrounded the rivets in the mast. The jib car tracks had corroded so badly that there was a layer of goop beneath them. Some of the bronze through-hulls had turned a frightening green, while a few of the seacocks were so corroded that they wouldn't budge. The stainless steel water tanks had rusted, too, and they leaked. Her appearance was at first so grim that I wished we had named her the *Unshine*, which would have been a very easy change from *Sunshine*. Instead we chose an obscure Greek word that nobody could pronounce or define.

But if *Syzygy* had cosmetic defects, we didn't care. Then we took her sailing. The diesel engine overheated on the way out of the marina, because the heat exchanger was caked up with rust. The reef hook had rusted so badly that it snapped the first time we furled the mainsail. Blocks had seized up, and the winches were so tight they offered little mechanical advantage. The wind vane almost fell off. Instruments didn't work, because the copper wires winding through the bilge had corroded so thoroughly that they no longer conducted electrical current. Shackles, turnbuckles, clevis pins, chain plates, backing plates, furler bearings, engine parts, the windlass axle—everything that could rust had rusted. Water, salt, air, and time had taken their standard toll, and corroded my bank account, too. That's how rust ate into my life . . .

RUST

INTRODUCTION:
THE PERVASIVE MENACE

Rust has knocked down bridges, killing dozens. It's killed at least a handful of people at nuclear power plants, nearly caused reactor meltdowns, and challenged those storing nuclear waste. At the height of the Cold War, it turned our most powerful nukes into duds. Dealing with it has shut down the nation's largest oil pipeline, bringing about negotiations with OPEC. It's rendered military jets and ships unfit for service, caused the crash of an F-16 and a Huey, and torn apart the fuselage of a commercial plane midflight. In the 1970s, it was implicated in a number of house fires, when, as copper prices shot up, electricians resorted to aluminum wires. More recently, in the "typhoid Mary of corrosion," furnaces in Virginia houses failed as a result of Chinese drywall that contained strontium sulfide. They rusted out in two years. One hundred fifty years after massive ten-inch cast iron guns attacked Fort Sumter, rust is counterattacking. Union forces have mobilized with marine-grade epoxy and humidity sensors. Rust slows down container ships before stopping them entirely by aiding in the untimely removal of their propellers. It causes hundreds of explosions in manholes, blows up washing machines, and launches water heaters through the roof, sky high. It clogs the nozzles of fire sprinkler heads: a double whammy for oxidation. It damages fuel tanks and then engines. It seizes up weapons, manhandles mufflers, destroys highway guardrails, and spreads like a cancer in concrete. It's opened up crypts.

Twenty-five miles northeast of San Francisco, one of the country's largest rust headaches bobs at anchor in Suisun Bay, and puts *Syzygy* to shame. Fittingly, the National Defense Reserve Fleet belongs to the US Department of Transportation, an agency that nearly plays God in its attempt to placate the needs of man and machine. Scores of people inspect on a daily basis as many old merchant ships that, in earlier extralegal times, would have been scuttled offshore. Now, the ships are too fragile to be hauled out and repainted, and not worth towing to Texas to be scrapped. Lacking other options, to Texas they've gone. Confounding matters, the US Coast Guard insisted in 2006 that the hulls of the ships be cleaned of invasive mussels before being moved, while the California Water Quality Control Board demanded that the bay not be polluted during said cleaning, and threatened to fine the Maritime Administration $25,000 a day until it came up with a plan. Environmental groups sued, demanding studies. While ten biologists, ecologists, toxicologists, statisticians, modelers, and mapping experts collected clams and mussels and took hundreds of sediment samples, the ships went on rusting. Big surprise: they contaminated the bay. At least twenty-one tons of lead, zinc, barium, copper, and other toxic metals have fallen off of the ships. What to do about the Reserve Fleet conundrum is such a touchy question that Senator Dianne Feinstein, who has a position on every environmental issue in California, officially has no position on the matter.

On the other coast, two dozen flip-flop-wearing employees of the US Naval Research Lab fill their time studying corrosion-resisting paints under palm trees at Naval Air Station Key West. Long before the place was an air station, in 1883, the Naval Advisory Board tested anticorrosive concoctions there, because rust was plaguing the navy. Today's paints self-heal, or can be applied underwater, or change color when exposed to rust—and still, rust plagues the navy. Rust, in fact, poses the number one threat to the most powerful navy on earth. By many measures, and according to many admirals (who sound as if they're employed by the DOT), the most powerful navy on earth is losing the fight. The name of one of the department's annual maintenance conferences: Mega Rust. The motto of that Florida lab: "In rust we trust."

As with boats, they say a lot of things about cars. About one brand of American car, they used to say this: "On a quiet night you can hear a Ford rust." In Ohio, since rust used to lighten automobiles by about ten pounds every year, that was half an ounce of metallic music to your ears nightly.* The symptoms extend beyond the rust belt, and express themselves in more than just Fords. Since 1972, the National Highway Traffic Safety Administration has had Volkswagen recall three-quarters of a million Sciroccos, Dashers, Rabbits, and Jettas with rusting fuel pumps and nearly as many cars with rusting brake lines. At NHTSA's insistence, Mazda recalled more than a million cars with rusting idler arms, and Honda recalled nearly a million vehicles with rusting frames. Chrysler recalled half a million cars with rusting front suspensions and Subaru recalled as many with rust problems in the other end. Ford recalled nearly a million Explorers with rusty hood latches and nearly a million Mercurys and Tauruses with rust-prone springs, and in the fifth-largest recall in history, almost four million SUVs and pickups because corroding cruise-control switches could cause parked vehicles to catch fire. You'd hear that day or night. Rust, attacking rocker panels, door hinges, door latches, floor pans, frames, fuel lines, airbag sensors, brakes, bearings, ball joints, shift cables, engine computers, and hydraulic hoses, has led to steering loss, wheel loss, shifting loss, fuel tank loss, brake failure, airbag failure, wiper failure, axle failure, engine failure, and hoods flying open at speed. DeLorean made its bodies out of stainless steel, old Land Rovers had galvanized chassis, and some 1965 Rolls-Royces had galvanized underbodies, but few automobile companies have steered clear of corrosion. Hyundai, Nissan, Jeep, Toyota, GM, Isuzu, Suzuki, Mercedes, Fiat, Peugeot, Lexus, and Cadillac have all recalled automobiles because of rust. More than once, Firestone has recalled millions of steel-belted radials on account of rust. Of NHTSA, the president of the consumer rights advocacy organization Public Citizen, Joan Claybrook, had this to say in 2003: "They've made up more names for recalls than Carter has liver pills." NHTSA never made up names for rust, though. It's always just corrosion. The godfather of American corrosion studies, a

---

*Because corrosion is exothermic, the skin of a corroding Ford becomes hotter than the metal underlying it, and this thermal gradient generates local stress called electrostriction. Technically, with the right tools, you really could hear it.

metallurgical engineer named Mars Fontana, once joked that in addition
to the eight forms of corrosion he had defined, an additional form was
"automobile corrosion."

In the twenty-one states that the DOT calls the "salt belt states"—the
upper right quadrant of the contiguous United States, everywhere north
and east of Kansas City, Missouri—it's not hard to suffer from the malady.
In postwar suburbia, state departments of transportation resorted to salt
(sodium chloride or calcium chloride) like addicts, doubling their use on
highways every five years until 1970. By then, the country used about ten
million tons of salt a year. It's fluctuated mildly since. Salt is bad news because chlorine is as reactive as oxygen, and more persistent. By 1990, the
total bill for nationwide salting was half a billion dollars; Robert Baboian,
a straight-talking corrosion engineer with much experience in public and
private consulting, contributed to a Transportation Research Board study
on the matter. No use in cutting back now, he wrote—the salt had long
since begun reacting with the steel in bridges, such that the chloride ions
were embedded like trillions of tics. Salting has much to do with the deficient condition of the country's bridges, but at least you can spin your
steel-belted radials on wet pavement on a snowy day. The cost of maintaining those bridges also had much to do with the DOT's funding of the
2001 study of the nationwide cost of corrosion, which made the cost of salt
look like peanuts.

Thanks to better design (eliminating areas that hold mud and moisture), galvanized parts, improved primers and paints, and tests in salt mist
facilities—giant steam ovens for cars—auto manufacturers got a handle on
corrosion more or less around Y2K. Bridges haven't caught up. As a result,
few other agencies are pulled in directions so opposite with such force as
the DOT. Yet there are limits to how far it may be stretched. A new car,
the agency figures, you can afford; a new plane, it figures, you cannot. At
airports, the Federal Aviation Administration prohibits the use of standard chloride-containing highway salts. Instead, airports rely on deicing
alternatives like acetates, formates, and urea. The most common, calcium
magnesium acetate, is one-fifth as corrosive as salt on steel and one-tenth
as corrosive on aluminum. It also costs twelve times as much as salt. To
deice planes, airports rely on glycols. If you really want your car to last,
drive exclusively down runways.

Beyond the domain of the FAA, rust troubles us almost everywhere. Oil rig designers put one extra inch of steel on the bottom of offshore oil platforms, calling it a "corrosion allowance." Some engineers mitigate "urine splash" in bathroom fixtures; others design bridges with corrosive pigeon poop in mind. More than a few engineers ensure that corrosion doesn't ruin your can of Coke before you get to it. Relying on corrosion tests (developed by Baboian), the US Mint designed new pennies and dollar coins. The government does not want, literally, to lose money. *Cloud Gate*, the sixty-foot, hundred-ton bean-like sculpture in Chicago, was made of a low-sulfur stainless steel so that it would remain shiny, and so that it would endure for a thousand years the road salt deposited by Chicago's *other* godlike agency. Engine oil, gasoline, and coolant all contain corrosion inhibitors—in concentrations from a few parts per million to a thousand times that. In gasoline, the inhibitors protect not just your car's fuel tank, but the gas station's underground storage tank, and the pipeline through which the gasoline was delivered. To protect water mains, tap water contains a corrosion inhibitor. Where I live, twenty-five miles east of the continental divide in Colorado, it's lime (calcium hydroxide), though other municipalities use sodium hydroxide or phosphates. Engineers in my town add the stuff from a fifty-thousand-pound tank, much like a flour sifter, to counteract the acidity that results from processing water. The clearest, safest, cleanest water just happens to be slightly acidic. As such, it's corrosive. They add lime to make the water slightly basic. As water flows from the Rockies to the Mississippi, and gets successively treated by more municipalities, it grows laden with calcium and magnesium, becoming what most people call hard. It's not like utilities are trying to make the water hard. They're trying to load it up with positive ions and make it less corrosive. City governments see water mains as the DOT sees planes: worth keeping operational as long as possible. Showerheads and faucets: clogged with minerals, they're as repairable or replaceable as Fords.

Only a small portion of Fortune 500 companies—those in finance, insurance, or banking—are privileged enough not to overtly deal with corrosion. Of course, corrosion is a major concern where their servers are stored. To inhibit rust in server rooms, companies use dehumidifiers and gas filters that remove ozone, hydrogen fluoride, hydrogen sulfide, chlorine, sulfur dioxide, and ammonia to minute (less than a few parts per billion) levels.

On the Principality of Sealand—a tiny platform in the North Sea—server rooms are filled with nitrogen, so that anyone entering needs to put on scuba gear. This anoxic environment provides a certain type of security, and ensures against corrosion.

Rust is so prevalent that regarding it the Bible offers a sense of defeatism. "Lay not up for yourselves treasures upon earth, where moth and rust doth corrupt, and where thieves break through and steal," says Matthew 6:19. Why improve thy lot if nature will unimprove your work while others plot to take it? A Yiddish proverb suggests the same inevitability: "Trouble is to man what rust is to iron." Rust has been such a mainstay that the British admiralty, in 1810, refused to hear a proposal on using iron rather than wood for ships. The Royal Navy figured that "iron doesn't swim." Lloyd's, too, wouldn't insure oceangoing metal ships for more than two decades after such things began moving cargo.

In industrializing America, where one author called rust "the great destroyer" and another simply referred to it as "the evil," corrosion seemed such a threat that urban critics considered it folly to build skyscrapers with steel. In Chicago in 1902, engineers debating corrosion rates predicted that the city's first steel structures would fall down in three years. In New York that same year, when one of the city's earliest skyscrapers, the eight-story Pabst Building, was eliminated (to make room for Adolph Ochs's twenty-five-story Times Tower), it was disassembled beam by beam, bolt by bolt, so that engineers could examine the effect of the damp climate on the steel. Many had said that erecting such buildings so near the coast was ridiculous.

By the end of the twentieth century, judicial opinion well recognized that rust was both inevitable and dangerous. Judge Linda Chezem, of the Indiana Court of Appeals, addressed corrosion in a case involving a leaking underground storage tank (UST) at a gas station halfway between Indianapolis and Chicago, smack in the middle of the rust belt. She wrote:

> Evidence was presented that Shell and Union understood that steel USTs are subject to corrosion; that steel USTs eventually leak; that leaks were impossible to prevent (prior to 1980s technology); that slow leaks were virtually impossible for the on-site gas station employees to detect (with the primitive dipstick method); that the solution to the problem required significant engineering knowledge and resources beyond the

limits of most gas station owners; and that small amounts of gasoline leaking into groundwater over a long time can pollute an entire community's drinking water with benzene, a human carcinogen.

In other words: *Oh, steel: that stuff's unreliable. Everybody knows that. Failure is destiny, and if we don't fix it, we'll all get cancer because of it.*

There's even rust in outer space, on account of atomic (rather than molecular) oxygen—no small challenge for NASA. Rust is ubiquitous. It's why cast-iron skillets are oiled, why copper wires are sheathed, why lightbulbs contain no oxygen, why spark plug electrodes are made of metals such as yttrium, iridium, platinum, or palladium, and why serious dental work costs an arm and a leg. The highest-ranked rust official in the country calls it "the pervasive menace."

Almost every metal is vulnerable to corrosion. Rust inflicts visible scars, turning calcium white, copper green, scandium pink, strontium yellow, terbium maroon, thallium blue, and thorium gray, then black. It's turned Mars red. On Earth, it gives the Grand Canyon, bricks, Mexican tile, and blood their hue. A ruthless enemy, it never sleeps, reminding us constantly that metals, just like us, are mortal. Were *Mad Men*'s Don Draper to pitch metal, he'd say it's like a maiden: rare, unrivaled in beauty, and impossibly alluring; but also demanding of constant attention, best watched carefully, quick to age, and intrinsically unfaithful. This of modern society's most important material!*

Yet rust sneaks below the radar. Because it's more sluggish than hurricanes, tornadoes, wildfires, blizzards, and floods, rust ranks dead last in drama. There's no rust channel. But rust is costlier than all other natural disasters combined, amounting to 3 percent of GDP, or $437 billion annually, more than the GDP of Sweden. That averages out to about $1500 per person every year. It's more if you live in Ohio, more if you own a boat like *Syzygy*, much more if you command an aircraft carrier.

---

*Oddly enough, iron retains a trustworthy metaphoric reputation: *an iron will, an iron fist, an iron hand, a mind like a steel trap.* As for the Man of Steel, who needs Kryptonite when saltwater will do the trick?

Nevertheless, rust is glossed over more than it's taught, because neither engineering students nor professors are drawn to it. It's just not sexy. John Scully, the editor of the journal *Corrosion*, told me corrosion gets no respect. "It's like saying you work in mold or something," he said. Ray Taylor, who runs the National Corrosion Center, an interdisciplinary agglomeration at Texas A&M that sounds bigger than it is, was more blunt. "We're sort of the wart on the ass of the pig," he said. A former rust industry executive said he and his colleagues always felt like the Rodney Dangerfields of the engineering community.

Sensing as much, we avoid the word. Residents of Rust, California, changed the town's name, a century ago, to El Cerrito. Politicians, too, know better than to mention rust. Though a few presidents have mentioned infrastructure and maintenance, none has mentioned corrosion or rust in a State of the Union address. President Obama has, between 2011 and 2013, called America's infrastructure failing, crumbling, aging, deteriorating, and deficient—but he didn't call it rusty. That's as close as a president has come to uttering the word. Like a condition between high cholesterol and hemorrhoids, rust is a nuisance that we'd prefer not to deal with, and certainly not talk about in public. Confidentially, industry representatives inquire with Luz Marina Calle, the director of the Corrosion Technology Laboratory at NASA's Kennedy Space Center, regarding their rust woes. Privately, Americans call John Carmona, the proprietor of the Rust Store, and ask for advice. Thanks to *New York Times* political columnist David Brooks, the threat of moral corrosion instills more fear than the threat of physical corrosion. But from those who ascribe no shame to talking about rust, stories emerge in the manner of scars and broken bones. People talk about the bottoms of their wells, their barbecue grills, their bicycle chains. They invoke a Neil Young line. Most often, stories begin with, "Oh man, I once had this car . . ." Was it, perhaps, a Ford?

As if to mask our reticence and apparent helplessness, most of us civilians fight rust like sailors. We attack with words. We're attacking metal with Rust Fighter, Rust Destroyer, Rust Killer, Rust Bandit; garrisoning with Rust Defender, Rust-Shield, Rust Guard; using weapons like Rust Bomb, RustBlast, Corrosion Grenade, and Rust Bullet—the latter of which is available in a Rapid Fire model or in a Six Shooter Combo Pack. The products suggest that in our reaction we're putting up a good fight.

But flight also works. Consider a 1960s newspaper ad for United Airlines. It said, "Prevent Rust. Periodic application of our Sunbird jets keeps the rust off your golf game, clubs, and you. United Sunbirds nonstop it to San Diego . . . Where the skies are not cloudy all day."

Golf clubs in San Diego still rust, as do ships at the naval base there. Jets in Tucson, Arizona, still rust, as do hammers in Oymyakon, Siberia, but five hundred times slower than they do at Punta Galeta, Panama.* This, according to measurements compiled by corrosion consultant extraordinaire Robert Baboian in the *NACE Corrosion Engineer's Reference Book*. Some fifteen thousand Americans working in corrosion have reason to consult this text. They range from linear and serious and introverted to scattered and rebellious and distractably sociable. Very few of them think of themselves as rust people. They work in "integrity management," or as coatings specialists, or as engineers or chemists. Reclusive or not, they're humble about their work. Many, I found, possess a keen awareness of their role in society, referring to themselves as members of the three "rusketeers" or of rust's three amigos. Because the world of rust is pretty small, most know each other. When tank cars full of chlorine spill, for example, the three amigos collaborate.

Most corrosion engineers are men. In my rough estimation, something like two thirds of these rust guys are mustachioed. I have a two-part theory about it. I suspect that (1) such men recognize that fighting the growth of hair on their upper lips is futile, and that trimming, combing, and maintaining makes much more sense; and that (2) such men, many of them technically minded engineers who work within strict bounds, have few other artistic outlets. Even Baboian, back in the seventies, had a mustache. They talk about galling, spalling, necking, and jacking; holidays, tubercles, and tubulars; pigs, squids, and perfect ends. About rust, one wrote a decent poem, but not one has yet devised a decent joke. Many have unique perspectives. "You're gonna be wrong a lot in corrosion," one, in Alaska, told me. "You're gonna think you have it pegged down, but you'll get nailed in the ass. It's a little adventure as you go along."

---

*Punta Galeta, Panama, wet six days a week, holds the world's highest corrosion rates for steel, zinc, and copper, and is conveniently located at the Caribbean entrance to the Panama Canal. For aluminum, though, the most threatening place in the world is Auby, France.

Fighting rust is more than adventurous. Often it's scandalous. For those trying to understand rust, prevent rust, detect rust, eliminate rust, yield to rust, find beauty in rust, capitalize on rust, raise awareness of rust, and teach about rust, work is riddled with scams, lawsuits, turf battles, and unwelcome oversight. Explosions, collisions, arrests, threats, and insults abound. So does war: rust and war have a long, tangled history, and together led to many fixtures of modern society. In addition to Big Auto, the following stories include Big Oil, whose products go toward Big Plastic and, in turn, Big Paint. A great deal of corrosion work entails studying how well paint sticks: to the Statue of Liberty, to the inside of a can, to the outside of a pipe, to the hull of a ship, to the hood of a Ford. The stories even include Hollywood and Big Tobacco. As the late senator Warren Magnuson of Washington, who on a warm March day in 1967 introduced the legislation that granted federal oversight of pipelines, would have said: they include all of the Big Boys.

Much of what follows explores our posture toward maintenance, and in that regard, reveals our humility (or lack thereof), willingness to compromise, and fundamental awareness. Rust represents the disordering of the modern, and it reveals many of our vices: greed, pride, arrogance, impatience, and sloth. It reveals the potency of our foresight, the weakness of our hubris, our grasp of risk, and our understanding of the role we fill in the world. What a predicament! Thus far, in the affairs of man and metal, our efforts have ranged from pathetic to ingenious, political to secretive. Of the many people I spoke to about rust, only one, an advisor to many federal science agencies named Alan Moghissi, saw rust as an opportunity. He imagined it could become as big as the environmental movement of the 1970s.

So easy to ignore, rust threatens our health, safety, security, environment, and future, and nearly got away with destroying our national symbol of liberty. Surrounded by stainless steel scissors, sinks, spoons, stoves, and escalator treads, we take nonrusting steel for granted—though it's only a century old. Expecting solutions, we disregard the management that most metals require. Can't we create a rust-free world?

A rust-free world would be a world without metal. In *The World*

Without Us, Alan Weisman cleverly illustrates the short-lived nature of metalworks. After only twenty years sans humans, he writes, unabated corrosion would destroy many of the train bridges on Manhattan's East Side; after a few hundred years all of New York's bridges would fail; a few thousand years from now the only intact structures would be those deep underground; and seven million years from now vestiges of Mount Rushmore would be all that's left to show that we'd once been here. Marc Reisner, in *Cadillac Desert: The American West and Its Disappearing Water,* writes that, much to our dismay, massive concrete dams—millions of cubic yards' worth—may be what we end up leaving for future archaeologists to ponder. I like to think that they'll consider those dams the same way we consider the Great Pyramid of Giza, the Great Pyramid of Cholula, the Great Wall of China, and the Parthenon. Better yet, I like to imagine them finding the granite foundation on Liberty Island, the statue above long since vanished, wondering, perhaps, if people once tried to dam the Hudson.

Like radioactive elements, most metals—the ones we rely on—have a half-life. But we don't recognize it. "We seem not at all resigned to the idea of major engineering structures having the same mortality as we," Henry Petroski writes in the classic *To Engineer Is Human.* "Somehow, as adults who forget their childhood, we expect our constructions to have evolved into monuments, not into mistakes. It is as if engineers, and nonengineers alike, being human, want their creations to be superhuman. And that may not seem to be an unrealistic aspiration, for the flesh and bone of steel and stone can seem immortal when compared with the likes of man."

If most of America's bridges, ships, cars, pipelines, and so on don't bring to mind mortality, surely one structure in the middle of Pittsburgh does. It's the U.S. Steel tower and eerily, almost menacingly, it looms above the city. It was built in 1970 of U.S. Steel's latest stuff: a "weathering steel" called Cor-Ten that works like stainless steel but develops a brown patina. A protective layer of rust. As the steel weathered, the brown developed on more than just the building. Embarrassing the hell out of U.S. Steel, the building stained the sidewalks below, until they had a distinct reddish tinge a block in all directions. The sidewalks have since been cleaned, and the building has since darkened into a hue best described as Darth Vader Lite. It looks dead. The psychology building at Cornell is made of the same stuff. Students at Big Red call it Old Rusty.

1

A HIGH-MAINTENANCE LADY

On Saturday, May 10, 1980, her caretaker slept in. David Moffitt awoke around eight o'clock and put on civilian clothes. He had a cup of coffee, then went out to the garden on the south side of his brick house, on Liberty Island, and started pulling weeds. A trained floriculturist who'd worked on Lady Bird Johnson's beautification efforts in Washington, DC, he had a spectacular vegetable garden. As the superintendent of the Statue of Liberty National Monument, he also had a spectacular backyard. As usual for a day off, he planned to do a little bit of gardening before going to Manhattan with his wife and three kids, to go shopping in the city or bike riding in Central Park. It was a clear day, about 50 degrees, with a steady light wind out of the southwest. Moffitt was on his knees, pruning roses, a couple of hours later, when Mike Tennent, his chief ranger, ran up and told him that two guys were climbing the outside of the statue. That was a first. Moffitt looked up, focused his hazel eyes, and confirmed the claim. So much for his day off.

It was about 150 yards from Moffitt's house to the statue, and on the walk there, he could hear visitors yelling from the base of the pedestal up at the climbers. "Assholes!" they yelled. "Faggots!" Their visits were being interrupted, and they objected, because they knew the situation was unlikely to end in their favor. Moffitt was already as mad as the visitors, but not for the same reason. He thought the climbers were desecrating the statue, and probably damaging it. Moffitt, who was forty-one, with thick dark brown hair and a Houston accent, had gotten the job—considered a hardship assignment on account of the isolation—because of his good track record with maintenance. The island, and the statue, had fallen into disrepair; the National Park Service recognized that its maintenance programs were wholly deficient. Moffitt was the first full-time caretaker in a dozen years.

Halfway to the statue, Moffitt stopped, and watched the climbers unfurl a banner. Liberty Was Framed, it said in bold red letters, above Free Geronimo Pratt. Until then, he'd figured the climbers were just pranksters. Now, though he didn't know who Geronimo Pratt was, he knew the duo were protesters. And he knew how to resolve the situation. The NYPD had a team skilled at removing people from high places—he'd seen footage on TV—and he would call them. So he turned around, walked to his office, and ordered the island evacuated. Inside the statue, an announcement blared over its PA system requesting that visitors proceed to the dock area due to an operational problem. In his office, Moffitt then called the National Park Service Regional Director's Office in Boston. He'd done this a few times before, and was destined to do it many more times again.

On his watch, Puerto Rican nationals had occupied the statue for most of a day, and a handful of Iranian students had chained themselves to the statue, protesting America's treatment of the Shah. On his watch, he dealt with about ten bomb threats a year. Before his time, the statue had been the site for college kids protesting President Richard Nixon, veterans protesting Vietnam, the American Revolutionary Students Brigade protesting the Iranian government, and the mayor of New York protesting the treatment of Soviet Jews. As Moffitt well recognized, the statue was the ideal place to protest any perceived wrong. So Moffitt called the NYPD, rather than the US Park Police—and this decision had ramifications for the climbers, and more importantly, the statue.

When the NYPD's Emergency Service Unit arrived, its agents were cheered by the departing visitors. They quickly assessed the situation. A "removal," they determined, would be too dangerous. They figured nets were needed. And helicopters. Given all of this, Moffitt figured that the situation might take a while to conclude and told his wife to go to

Manhattan without him. Then he learned from the NYPD that Geronimo Pratt was a Black Panther convicted in the murder of a Santa Monica teacher, a crime for which he'd been imprisoned for a decade, and he remained angry. There was nothing admirable about desecrating the statue, no matter the cause. "I took the job of protecting this symbol of America very seriously," Moffitt recalled.

Moffitt spent the day in his office, watching the climbers through a pair of government-issued binoculars. That afternoon, he took a call from a reporter at the New York *Daily News*. In the middle of the interview, he heard a banging sound coming from the statue. "God damn them!" someone below the statue yelled at the same time. "They're busting my statue!" A ranger came in the office and said one of the climbers was driving pitons into the copper skin. Moffitt doesn't recall how many bangs he heard, but he remembers being frantic. Now he was sure they were damaging his statue. He yelled at the reporter, then hung up.

Up on the statue, Ed Drummond, a thirty-four-year-old English poet from San Francisco with an arrest record for climbing buildings and hanging banners, was struggling. After traversing around the left foot, then up and left, the climbing became more difficult than he had expected, or had been prepared for. It had taken him two hours to get to the crook of Lady Liberty's right knee, and now he was stuck on a small ledge, looking up at a short chimney in the folds of the robe on her back. The surface of the copper skin, in particular, was causing problems, rendering his two eight-inch suction cups useless. The skin was covered in millions of little bumps, almost like acne, the result of the French craftsmen who pounded the copper into shape a century before. Consequently, his suction cups stuck only for about ten seconds, even if he pushed with all of his might. "I realized that they were not going to work," he recalled, describing the fatigue he began to feel in his arms. He slipped, slithered down a few feet, and barely caught himself with his other suction cup. He was aware of the consequences of falling. "You'd just go hurtling out into the air," he recalled, "and end up two hundred feet down on the esplanade." It was also almost certain that if that happened, he would pull his climbing partner, Stephen Rutherford—a thirty-one-year-old teacher-in-training from Berkeley, California—off too.

As he climbed, he could see that between the plates of copper there was often a small gap. The plates had begun to lift for some reason, though the edge formed was not big enough to use for climbing. He also noticed many little holes in the statue, which he had not seen from the ground. Rumor had it, among Statue of Liberty buffs, that they were bullet holes. As the climbing grew more desperate, with his back on one wall of the chimney and both of his feet on the other, he tried placing a tiny S-hook, which he'd bought last minute, in one of the holes, for support. Using a sling, he weighted it, and under less than his full weight, it bent alarmingly.

Drummond had planned to climb up the statue's back, and onto her left shoulder, then stay in a little cave under the lock of hair over her left ear. Sheltered from wind and rain, anchored to that lock of hair, he planned to keep a weeklong vigil. (He brought a sleeping bag, and a supply of cheese, dates, apples, canned salmon, and water bottles.) He planned to drape his banner across the statue's chest, like a bra. But he never made it past the chimney. Instead, he decided to spend the night on the ledge, and descend in the morning. He told as much to the NYPD, who relayed the information to Moffitt. That night, Moffitt didn't get much sleep. From his bed, through his window, he watched Drummond and Rutherford. His children complained about all of the hubbub and helicopters flying around.

The next morning—Mother's Day—Drummond and Rutherford surrendered, more or less twenty-four hours after they'd started. By the time they'd rappelled to the statue's feet, the press had shown up on the mezzanine. A reporter yelled up, "Did you use any pitons?" Immediately, Drummond yelled down, "No, we haven't damaged the statue!" Then, below the small overhang formed by the little toe of the statue's left foot, he yelled, "This is how we climbed the statue!" and pressed one of the suction cups against the metal. He and Rutherford hung from it. As they descended into the scrum of police waiting with handcuffs, Drummond insisted, again, that he hadn't damaged the statue. Moffitt, though, later told the Associated Press reporter that the climbers were "driving small spikes" into the statue. As he was talking to reporters, someone handed Moffitt a note from the US Attorney's Office. It said, "Do not offer them amnesty." Moffitt wasn't about to. He was furious.

After a night in jail, Drummond and Rutherford were charged with criminal trespassing and damaging government property, to the tune of $80,000. By then, Moffitt had studied the statue through his binoculars, and discovered the same holes that Drummond had. He'd also sent one of his maintenance guys up the statue to inspect the damage from the inside. He discovered that the holes were everywhere, and weren't the result of pounding pitons, or spikes of any kind, into the copper. They were places where the rivets, which held the statue's copper skin to her iron frame, had popped out. The holes in the statue hadn't been created by Drummond at all. They'd been created by corrosion.

So Ed Drummond was right. Liberty was framed, and her frame was rusting.

What had been interpreted as an act of vandalism turned out to be a much bigger headache for Moffitt. Sure, there was graffiti on the inside of the statue, but nobody had ever damaged the outside. At least Moffitt was pretty sure. Perplexed, he dug through a file cabinet, in search of reports on the statue's condition. He found none. So he had a short section of scaffolding put up to inspect the damage on the statue. Scuff marks, and small spots where Drummond's rope had worn away the green patina, were discovered. He also called the National Park Service's design/construction firm in Denver, and asked their engineers to examine the statue and report on its condition. Two engineers came out a few weeks later, and investigated. They wrote a memo, and gave it to Moffitt. It concluded that the statue was basically sound, corrosion notwithstanding, and did not recommend any repairs. Moffitt was relieved they'd found no damage, but was disappointed that the inspection was solely visual. He was hoping for something more. He'd seen the damage himself, and he wanted answers. So on May 20 Moffitt had two of his staff ask the Winterthur Museum, which had examined the Liberty Bell, to determine the "causes of the severe corrosion and make suggestions to stabilize the system to avoid catastrophic destruction." They sent two copper samples from Lady Liberty's torch to the museum, and the museum put them in front of Norman Nielsen, a metallurgist at DuPont.

Nielsen's report wasn't much more illuminating than the one from Denver. "It was hoped," he wrote, "that such a study would define the corrosion processes that appear to be causing the copper to deteriorate at an alarming rate and which might suggest measures that might be taken to stabilize the corrosion process." Instead, his investigation, achieved via

X-ray fluorescence, merely identified the chemical composition of the copper, its patina, and some of the impurities within, including antimony, lead, silver, zinc, and mercury.

Two days before Nielsen finished his report, Drummond's case was heard. It was obvious, by then, that Drummond hadn't put the holes in the statue, and the damage charges were dismissed. After all, Drummond had brought no pitons, and no hammer—as was recorded in the report of his arrest, during which his backpack was searched. The banging sound, it turned out, had come from a police officer rapping the butt of his gun on the inside of the statue. But Drummond was convicted of trespassing, a misdemeanor for which he was sentenced to six months of probation and twenty-four hours of community service.

A few months later, Moffitt received a phone call from a lawyer representing a couple of French engineers who'd just completed the restoration of a similar copper and iron statue, called Vercingetorix. They offered to do a more thorough investigation of the Statue of Liberty, which, after all, had been a gift from France. (It's not surprising that France beat us to the punch; France's history with metal structures is generations older than America's.) Moffitt was all for it, as his questions remained unanswered, and he knew further inquiry would be limited by severely restricted NPS funds. The coincidence was serendipitous, to say the least, as Moffitt had twice suggested the formation of a commission to plan for the statue's upcoming hundredth anniversary, but gotten nowhere, because of budget constraints under President Jimmy Carter. He knew the statue would need to be spiffed up, but nobody, it seemed, wanted to hear about it, much less pay for it. So Moffitt met the engineers at Liberty Island, and arranged for them to meet with the director of the Park Service. A year after Drummond's ascent, they agreed to form a partnership to restore the Statue of Liberty. The years of "neglect and deterioration," as the Park Service referred to the 1960s and 1970s at the statue, were about to end. Amazingly, what had begun as an obscure attention-getting stunt by two protesters ended with the most symbolic rust battle in this nation's history.

The rusting statue—once the world's tallest iron structure—was a mystery. As seven architects and engineers from France and America began to research her past, they pieced together details. What was clear was that she had been managed, or mismanaged, in a variety of ways, by a mess of agencies. She'd been built in 1886, on top of Fort Wood, on Bedloe's Island, and after two weeks of orphanage, was initially overseen by the US Light-House Board, which was part of the Treasury Department. She spent fifteen years in that agency's care, and then twenty-three years under the War Department, before she was declared a national monument. Nine years later she was transferred to the National Park Service. In other words, a half century transpired before anyone with a sense of preservation took over caring for her. One of the first things the NPS did, with the Works Progress Administration, in 1937, was replace parts of her corroded iron frame. Good preservationists, they replaced iron bars with similar iron bars. But, because all of the work was done from the inside of the statue, they used self-tapping screws, rather than rivets. You could say they botched the job. Since then, the statue hadn't received much better care; the monument hadn't had an official superintendent since August 1964. There'd been a management assistant, three assistant superintendents, one acting assistant superintendent, a unit manager (none for more than two and a half years), and finally Moffitt, in January 1977.

The American half of the team—Richard Hayden, Thierry Despont, and Edward Cohen—wanted more detail about the statue's past, so they visited other statues built by the statue's architect, Frédéric-Auguste Bartholdi, and its engineer, Alexandre-Gustave Eiffel. They went to the Bartholdi museum, in Colmar, France, to see notes, papers, models, and a journal from 1885. They found no drawings, but found out that Bartholdi never intended visitors to see the inside of the statue, which complicated things, because Americans loved that part of her. Elsewhere, they found Eiffel's sketches, and nine handwritten pages from November 12, 1881, showing calculations for the statue's frame—explaining how 270,000 pounds of iron could support 160,000 pounds of copper.

The frame's design—an iron skeleton riveted to the copper skin—was ingenious, and risky, and Bartholdi had known it. In fact, he'd originally chosen another design, by Eugène Emmanuel Viollet-le-Duc, in which the statue was filled up to the hips with sand. But Viollet-le-Duc died in 1879, so Bartholdi went with Eiffel. Eiffel's design was risky because the two metals couldn't actually touch each other. Dissimilar metals, in contact, would corrode, as Luigi Galvani had discovered a century before. The corrosion had a name: galvanic corrosion. It's how batteries work, actually. Electrons travel from the weaker, more electronegative, metal, to the stronger one—and in the process the weaker one is destroyed, which is why batteries don't last forever. In the case of the Statue of Liberty, the voltage was only about a quarter of a volt—not enough to illuminate even the smallest lightbulb—but persistent, far more so than any battery. Eiffel was aware of the risk, and planned to manage it by separating the iron from the copper with shellac-impregnated asbestos. It was the best technology of the era, and he had faith. "In regard to the preservation of the work," he wrote, "since all the elements of its construction are everywhere visible on the inside in all their details, it will be easily kept in good condition." *Scientific American* saw it differently, and within a month of the statue's completion, warned, "There are five dangers to be feared, namely, earthquake, wind, lightning, galvanic action, and man." Bartholdi took up the defense. "I have no doubt that with care and looking after, the monument will last as long as those built by the Egyptians," he wrote, after the statue was built. Things turned out differently, partly because he had never planned for paint.

It's not clear who first decided to paint the interior of the statue, but the job was done thoroughly in 1911, with a layer of black coal tar. On top of that, some other copycat in 1932 slathered a layer of aluminum paint, and someone else, in 1947 added enamel paint, specially formulated for removing graffiti. Before Moffitt arrived, at least six others ordered a layer of paint thrown on top of the others, for good measure. Intent on preserving the statue, Moffitt followed their lead. One of the first things he did as superintendent was paint the inside of the statue, with a light-green lead-based paint. Where there should have been a sign that said Caution: High Corrosion Risk, there was just layers of paint. All of that paint was almost as thick as the copper, and unfortunately, had trapped water between the iron frame and the copper skin—exactly what Eiffel and Bartholdi had wanted to avoid. Water between the copper and iron was as bad as having the two metals in contact with each other. Hence one of the American team's first discoveries: the statue had become an enormous battery. As a result, corrosion had produced a lot of "wastage," and in some places, paint was the only material holding things together.

The French half of the team, meanwhile, began collecting scientific, rather than historic, data. They installed wind gauges on the outside of the statue, and 142 strain and acceleration gauges in it. They installed carbon dioxide and humidity gauges inside, too, to measure condensation from the breath of millions of visitors in an enclosed place that, in the summer, regularly climbed above 120 degrees. They X-rayed the frame to check for cracks, and at Cetim, the Technical Center for the Mechanical Industry, in Senlis, France, they ran fatigue and impact tests on samples of the puddled iron frame, to see how cracks formed and propagated, and how the metal reacted to dynamic stresses caused by wind. They used ultrasonic calipers to measure the thickness of the copper skin, and photographed every one of the three hundred copper plates.

By December 1981, the French-American Committee for the Restoration of the Statue of Liberty had produced a preliminary diagnosis report, confirming Moffitt's suspicions that the statue was not, as Denver had declared, "basically sound." On July 14, 1983, the French-American Committee published a thirty-six-page magazine-like report, offering four restoration proposals. The proposals varied only in the extent to which they would improve accessibility and conditions for visitors: stairs, elevators, resting platforms. Otherwise, the proposals included the same amount of structural repair, "to assure the integrity of the structure and avoid additional electrolysis"—by which they meant corrosion—"for the foreseeable future."

Everywhere engineers had looked, in every part of the statue, they found corrosion, or a contributor to it. Only the exterior of the copper skin had withstood corrosion, and been deemed "normal," rivet holes and other damage notwithstanding. Once the iron frame had begun to rust, the degradation spiraled out of control. When a spot on the frame rusted, it swelled, and inhibited movement of the flexible joint between the copper skin and the iron frame (which was there to allow for the slight expansion and contraction of the copper), which then warped the copper, and eventually pulled rivets out, putting yet more strain on the copper skin. It was called "jacking," and it was like a chain reaction. More popped rivets meant more water getting in, especially because of the pressure difference between the inside and the outside of the statue. Lady Liberty almost sucked water in. One-third of the statue's twelve thousand framing rivets were loose, damaged, or missing, and more or less half of the frame had corroded. The asbestos insulator—which actually wicked water, exacerbating the damage—had long since disintegrated. As a result, some of the ribs in the frame had lost two-thirds of their thickness. The lattice girders below the statue's robe and feet were "particularly corroded"; in photos, it looks like some sort of metal beaver chewed away at them. The frame of the book in her left hand was "very corroded," and beneath her crown it wasn't much better. The staircase was rusted badly. The corrosion in the frame of her right arm was severe, in the torch it was "extensive." Corrosion in the whole frame was so "deleterious" that the system was said not to function anymore. There was, according to the report, "definite risk of structural failure" in the torch, an event that would be embarrassing to say the least.

Water was entering the statue through the rivet holes, through badly designed weep holes (intended to let water *out* of the statue), from the lungs of millions of visitors, whose breath condensed inside the statue, and from a sizeable hole in the statue's raised bicep, where one of her seven spikes was poking through. The statue's lack of watertightness was most obvious in the winter, when it was easy to find snow inside the statue. Water was also coming in from the torch, which had been a disaster from the beginning.

In 1886, as the statue was assembled—or reassembled—in America, Bartholdi had wanted eight lights on the torch to illuminate the gilded copper flame. A week before the statue's inauguration, on October 28, the US Army Corps of Engineers told him that the lights would interfere with boat navigation in the harbor, and that his design would have to be modified. Lieutenant John Millis, of the US Light-House Board, decided to cut two rows of portholes in the torch, and illuminate it from within. The illumination was pathetic, barely visible from Manhattan. Bartholdi said the flame was like the "light of a glow worm." In 1892 the upper row of portholes was enlarged into an eighteen-inch band of glass, above which a skylight was added. Bartholdi remained unsatisfied. Twenty-four years later, a dozen years after he died, an artist named John Gutzon Borglum attacked the torch, sculpting away much of it. He cut out 250 rectangles, and put in 250 panes of amber glass. Borglum went on to attack Mount Rushmore. A metalworker later wrote that the torch resembled "a shapeless Chinese lantern," though it also resembled a huge bird cage, inside and out. The windows leaked, and the ventilation holes below were perfect bird entrances. Hence all of the rust.

The torch—the highest, wettest, windiest, least inspected part of the statue—was also the most delicate part. It had been made of thinner metal, to allow for intricate details on the soffit above the handle, and the pendant below. Up above the Hudson, it was also the most desirable spot for birds to roost. As a result, it was the most damaged part of the statue. Early on in the restoration, Hayden and Despont, of the American restoration team, climbed into the torch with a few park rangers, to check it out. In the bottom of the torch pendant, there was a stagnant puddle of water and bird poop, which they called a "primordial soup." The mixture was eating through the metal. If not for a threaded rod, with a large bolt, the pendant would have fallen off. They snapped a photo of themselves up there. At the next meeting, they passed it around. They were promptly advised by other engineers not to attempt that "daring feat" again, because the frame in the torch was seriously weakened. The frame, in fact, was missing. There was only a shadow of where it had been.

As the scale of the restoration became evident, the French-American Committee was superseded by an American commission and foundation for the restoration of the Statue of Liberty. The organizations raised money, investigated, prepared, and finally got around to fixing the Statue of Liberty. Since the research and planning alone had taken three years, the next three years of restoration were hectic. They formed subcommittees, coordinating committees, subgroups, advisory groups, and *groupements* before offshoot state-level commissions and foundations latched on. They held meetings at the Waldorf Astoria, and they took flights to Paris, for walks at Versailles. The work was accomplished by more than three hundred workers—consultants, *compagnons*, experts of all kinds—working for more than thirty contractors. Prominent companies helped with technical research, and so many companies offered to donate tools and materials that hundreds were turned down. Even NASA chipped in. The foundation ran the largest direct-mail campaign ever, and eventually the most successful fund-raising campaign in American history. So that they could get materials to the island, they repaired a pier, then built a 1,200-foot bridge, from New Jersey to Ellis Island, because it was cheaper than transporting supplies on barges. Around the statue, they erected the world's tallest freestanding scaffolding, and eventually they fixed the statue up properly, drastically increasing her life expectancy. It was all overseen by Lee

Iacocca—the man who saved Chrysler—who was appointed chairman of the endeavor by President Ronald Reagan on May 17, 1982.

Iacocca said he'd raise $230 million, $300 million, $500 million, or $1 billion if he had to. The fund-raising effort began in New York, and soon spread to Los Angeles, Chicago, Atlanta, and Dallas, where fund-raising offices were opened. A gala at New York's Lincoln Center, with Luciano Pavarotti and Bob Hope, raised $750,000. Gerald Ford appeared at an event in Tennessee, pulling in less money. A toll-free 800 number was purchased, for phone pledges. Congress authorized the minting of thirty-five million commemorative coins. American Express donated a portion of all Traveler's Check sales.

Schoolkids all over the country collected pennies, sold muffins, and grew flowers for the cause. By July 4, 1986, kids at more than twenty thousand schools had raised over $5 million. A disabled six-year-old in Indianapolis raised $3,000. Ethnic organizations pledged money: Italian groups, Czechoslovakian groups, Greek, Polish, Serbian, Byelorussian groups. The Daughters of the American Revolution raised $500,000. A disabled-veterans group raised $1 million. Former employees of Bell Telephone raised $3 million. Employees from the State Farm Insurance Company raised $1 million. So did employees at Chrysler. Los Angeles threw in $50,000.

The post office printed commemorative twenty-two-cent stamps, introducing them at New York's Federal Hall, where George Washington had taken the nation's first presidential oath. To mark the occasion, the USPS arranged to remove forty pounds of copper from the statue, and had it melted and formed into two fifteen-inch replicas. The replicas were shipped to Cape Canaveral, loaded onto the space shuttle *Discovery*, removed from the land of liberty and the land of gravity, and then shipped back to New York, where one was melted down yet again into Official Centennial Seals, and the other now resides in the museum in the pedestal of the statue.

Ultimately, Iacocca's campaign raised $277 million ($1.4 billion in today's dollars) and threw it at a three-hundred-foot-tall metal object on an island on the windy, rainy, salty, humid Atlantic Coast.

⌒

It took three months, and $2 million, to erect the scaffolding around the Statue of Liberty. Engineers had considered bamboo scaffolding, Asian style, and also considered a pyramid. They considered a lattice anchored by cables, like a suspension bridge. They settled on a grid of all new aluminum poles, coated in zinc, to prevent the metal from staining the copper. It weighed three hundred tons and used two miles of half-inch steel cable, and was strong enough to support the statue's right arm, and able to withstand winds of up to one hundred miles per hour. By April 1984, you could climb up, lean forward, and give the statue a kiss.

On July 4, the old torch was removed and lowered to the ground. Seven months later, it was on the lead float in the Tournament of Roses Parade, with Miss America. It had been escorted to the airport by the NYPD, and shipped to California in a special container, then guarded by Park Service rangers. No rusty bird cage has ever been treated so well.

Workers on the scaffolding got a close look at the exterior of the statue, and discovered many surprises that received less fanfare. They found graffiti: a *B* from Bartholdi, some names of the men who'd worked on the statue in 1937, and, on the big toe, the inscription "Alone with God and the Statue, Christmas Eve." Drummond had left no John Hancock. They found bird nests in the folds of the statue's robe, with masses of guano dating from the nineteenth century. This they scraped away. They found torn rivets that looked like dimples, and tears in the copper, and stains where coal tar and paint had oozed through the seams. There were paint spatters on the back of the statue's arm, and a dark splotch on her back, either from Liberty Island's trash incinerator, or from acid rain. Significant pieces at the bottom of the statue's hair curls were missing, having corroded away. She had scabs, symptoms of "bronze plague," on her crown rays. There were cracks in her left eye, in her lips, in her nose, and in her chin. She had a big stain on the front of her neck, almost like drool. She had rust boogers. The condition of her skin was so bad that the French-American Committee proposed coating all of it in clear resin, a suggestion that Bartholdi had made ninety-four years earlier. Instead, damaged sections—almost 2 percent of the statue's surface area—were fixed as required with new copper. The hole in her bicep was patched, and her spike readjusted a few degrees.

Inside the statue, repairs were more difficult. The lead paint, coal tar, and disintegrated asbestos had to be removed before anything could be done to the copper. Men in white suits spent two weeks removing the paint by freezing it off with liquid nitrogen. Once frozen, it flaked off in sheets. Union Carbide showed them how to do it, waving magic wands. It took three weeks, and 3,500 gallons of liquid nitrogen. To remove the paint from the iron frame, a company called Blast and Vac was summoned. It invented what looks like a giant electric toothbrush. "Basically," a company rep described, "we installed a standard blasting nozzle inside a vacuum cleaner head." But still, the coal tar remained. The coal tar was more stubborn, reacting as it had with various corrosion products. Sandblasting would have removed it, but also would have damaged the copper, which was only $\frac{3}{32}$ of an inch thick. Same for most other abrasives, and most solvents. Engineers tried blasting samples of copper with cherry pits, ground corncobs, plastic pellets, walnut shells, powdered glass, salt, rice, and sugar, to no avail. Finally, without consulting the Park Service's corrosion consultant, none other than Robert Baboian, the Park Service settled on sodium bicarbonate: baking soda. Arm & Hammer reassured the NPS that it wouldn't damage the copper, and donated forty tons of it.

Since late 1983, Baboian had visited the statue dozens of times; climbed all over her. By the time he showed up in January 1985, baking soda was caked inches thick inside the statue and was leaking out through many holes. Workers had blasted it at 60 psi (pounds per square inch) all over the coal tar. On the inside, it was reacting with the copper, and turning it blue; on the outside, it was destroying the patina, and dramatically staining the statue in places. It was a disaster, even if it removed the coal tar. "It was a big mistake," Baboian told me. "It did a job on the outside of the statue." Arm & Hammer claimed the baking soda "did not harm the copper" but admitted there were some "unexpected results." Immediately, the inside of the statue got a wash with mild vinegar, and the outside received a daily shower until the baking soda was gone. The blue tint went away in a few weeks, but the stains remain, because it takes about thirty years for the patina to fully form.

Baboian knew a lot about the patina. Wisely, he compared the thickness of the exposed copper to a spot where some of the black coal tar had oozed out and covered it, thus protecting it from both sides, and determined the rate at which the copper was corroding. It was vanishing at a rate of .0013 millimeters per year. At that pace, he figured it'd last a thousand years. The patina on the dark patches on the statue, Baboian found, were a mineral called antlerite, rather than brochantite, and it was anywhere from one half to one tenth as thick. Thomas Graedel and John Franey, at AT&T's Bell Laboratories, in New Jersey, furthered his research, and took nine samples of copper from the statue, and seven samples from similar copper roofs at their office, and examined the patina's growth, depth, formation, cementation, and erosion. Using mass spectrometry and X-ray diffraction, they found chlorine trapped inside the patina, which was bad news. Chlorine has an insatiable appetite for metals. But they also determined that the patina didn't erode as long as the pH of rain or fog was above 2.5. In fact, they found that patina growth was twice as rapid as it had been a century before. They also matched a sample of the copper in the statue with a sample of copper from the Visnes mine, on Norway's Karmoy Island, and put to rest the debate about the source of her metal.

Baboian (who had run Texas Instruments's corrosion laboratory) also studied the interaction between the copper and the iron frame. The statue, he determined, "was an ideal configuration for galvanic corrosion." On account of the copper, the iron was corroding one hundred times faster than it would by itself. Worse, because the surface area of the copper was so large compared with that of the iron, corrosion was sped up another tenfold. The interaction also retarded the corrosion of the copper, which is what formed the green patina. As Martha Goodway, a historian at the Smithsonian Institution, later wrote, "the structural design of the statue was innovative, but the materials chosen to realize this design were not." Had the statue been built only a decade later, steel, rather than wrought iron, would have been used, and the story would be different.

So Baboian and others set about determining what type of metal to replace the iron with. In March 1984 he began testing five different alloys on a beach in North Carolina, at the LaQue Center for Corrosion Technology. Since the replacement metal had to have similar properties to iron, and be compatible with copper, he had few practical choices. He tested a plain steel, an aluminum-bronze, a copper-nickel alloy, a new alloy called ferralium, and marine-grade stainless steel, which was invented a generation after Bartholdi decided to employ iron in the statue. Because the samples were eighty feet from the shore, they corroded twenty-two times faster than they did inside the statue. After six months, Baboian had the equivalent of eleven years of corrosion to examine, and ruled out all but the ferralium and the stainless steel, which he recommended to the Park

Service. Engineers then figured out that ferralium wouldn't work, because bending it required heating it, and heating it destroyed its anticorrosive properties. Stainless steel it was.

Repairing the statue's frame proved the most challenging task yet. The frame was in such bad condition that the notion of preservation was abandoned. The whole thing was replaced, piece by piece. That meant 1,825 unique six-foot ribs, weighing 25 pounds each, and all of the fasteners necessary to connect them to the copper: a couple of thousand U-shaped clamps, nearly four thousand bolts, and twelve thousand copper rivets. The rivets were prepatined, so that it wouldn't appear that the Statue of Liberty had chicken pox. Because the frame was already weak, it was necessary to distribute the work, so as not to overstress the structure. The statue was divided into quadrants, from which one rib was removed at a time. Once removed, the copper was braced. The rib was brought to the metalworking shop near the base of the statue, where the torch was also being remade. There metalworkers fabricated a new rib, exactly like the old rib, by running 30,000 amps through it for five minutes, until, at 1,900 degrees, it became bendable. Once bent to the proper shape, it was quenched in water, sandblasted, labeled, wrapped up, and sent to Manhattan, where it was treated with nitric acid to re-create the outer, corrosion-resistant, patina-like layer of the metal. From removal to replacement, the process took thirty-six hours, including one hour just to manhandle the new rib back up the confines of the statue. The metalworkers worked around the clock for six months, fabricating seventy ribs a week.

Where the asbestos had been, isolating the skin from the frame, workers now used Teflon tape. Where Drummond had found gaps between the copper plates, workers now sealed the seams with silicone, kitchen style. To keep condensation from forming, a humidity-control system was installed. On the frame's main girders, workers slathered three coats of an inorganic zinc paint that had been developed by NASA, and tested in Hawaii; Astoria, Oregon; and on San Francisco's Golden Gate Bridge. (Because zinc is less noble than iron, it would corrode, rather than the frame, just as the iron had corroded when partnered with copper.) The manufacturer created a water-based, rather than solvent-based, version of the paint, so that fumes wouldn't turn the statue into a giant gas tank. On top of the paint, they put a layer of epoxy, to make graffiti-removal easier. By the time engineers got to the statue's shoulder, the preservationists had won out. The lady's shoulder was offset by a foot and a half, and her head was offset by two feet—an assembly error. Even though the frame there was overstressed and overly flexible, engineers decided to brace it rather than rebuild it.

Finally, on July 4, 1986, a new torch was lifted into place. The torch had been meticulously designed to Bartholdi's original plan: a solid flame, to be illuminated by lights from the outside. Instead of just copper, though, the flame was covered in gold leaf. Beneath it, the copper plates, which had been joined with 2,600 rivets, which were filled with solder and ground flush, were degreased, then chemically etched, then primed, then covered in three layers of varnish, the recipe for which dated from the eighteenth century, as on a violin. The gold was applied while the last coat was still tacky. Over the vents in the pendant below, bird screens were installed.

Take that, rust.

Reagan hailed the Statue of Liberty restoration as one of the highlights of his presidency, and as a flagship initiative on public-private partnership. Private companies had donated generators, cranes, paint, and copper, and hundreds of thousands of hours of their engineers' time. Black & Decker donated tools, and replacement tools. John Deere donated tractors. Coca-Cola lent $500,000, interest free. Sealand donated shipping containers for storage. The Cabot Corporation donated ferralium, the Specialty Steel Institute of North America donated stainless steel, AT&T donated its prepatinated roof, and Arm & Hammer might have been wise baking a few million muffins. The Ad Council donated $50 million of airtime—the most ever allotted to one campaign. Yet some corporate relationships raised hackles. Some company executives, asked to donate materials, stipulated plaques signed by Reagan thanking them for their work. One marketer proposed removing the statue's arm solely for publicity and admitted that he aimed to get rich on the restoration. He asked Moffitt if he wanted to resign and join his board, so he could get rich too. Another marketing group, actually hired, scammed the restoration foundation, proceeding on promotional projects—like twenty-five million boxes of Kellogg's cereal—without approval. Objectionably, the head of the group garnered

separate marketing contracts from statue sponsors including the Chateau
Ste. Michelle winery, U.S. Tobacco, and *Time* magazine. The Department
of the Interior questioned his scruples. Another company offered $12 mil-
lion for parts and materials removed from the statue, to be fashioned into
gewgaws—a proposal that even the Park Service found undignified.

The commercialization irked many. Michael Kinsley, in the *New Re-
public*, feared that Miss Liberty was becoming a "high-priced tart." The
New York Times editorial page warned that "no restoration is worth putting
a national monument on the market." Richard Cohen, in the *Washington
Post*, wrote that the price for the statue's survival "should not be her virtue." George Will saw nothing wrong. Commercialization was described
in newspapers as "an insult to Emma Lazarus's tired, poor and huddled
masses," and as "cultural sacrilege." Even media rights became an issue, as
the centennial approached. How could ABC purchase exclusive rights—
for $10 million—to a celebration, on public land, of a national holiday?

That was the least of the difficulties. Philip Kleiner, of Lehrer McGov-
ern, the construction management company, described the restoration as
the "ultimate fast track project," because every job was unique; most contractors weren't willing to get involved, and those that thought they could
do the work usually couldn't. The chronicler of the restoration, Ross Hol-
land, wrote that, "if Murphy's Law ever applied to anything, it applied in
spades to the Statue of Liberty restoration project."

There were long delays, attributed to the French team, which seemed to
be in no rush, and was abandoned by the Americans in August 1984. After
that, the pace was hectic; construction managers, who complained about
doing research and development at the same time, got their final drawings
delivered on July 3, the day before the torch returned.

There were cost quibbles, after millions of dollars were wasted on a year
and a half of nothing much, and another couple million was wasted on a
TV special. To make up for it, an executive vice president suggested cost
cutting by using plastic rather than ceramic tiles, and regular steel rather
than stainless. American companies scrutinized contracts that went to foreign companies, claiming they could do the same work for a third less, and
employ New Yorkers too.

But even when the work went to New Yorkers, unions bickered over
turf. Though Liberty Island belongs to New York, the piers and docks

belong to New Jersey, and Congressman Frank Guarini, along with the mayor of Jersey City, pushed for at least half of the labor to come from New Jersey unions. When the scaffolding was to be delivered, via boat, by a member of the carpenters' union, the marine operating engineers' union complained and threatened to picket the Marine Inspection Office, clogging up works at the Coast Guard. Teamsters and the International Brotherhood of Electrical Workers battled over the electrical contract. Union 580, a New York ironworkers union, protested the granting of the torch-building to a French firm—and picketed the press conference announcing their selection. Not quite Drummond-style, they unfurled a banner, then gave the French workers hats and T-shirts featuring the number 580.

The political battling was uglier. After Iacocca denigrated Reagan's economic policies, in the spring of 1984, rumors began circulating that he was considering a run for president as a Democrat. On Independence Day that year, when the old torch came down, Reagan snubbed Iacocca by not attending the ceremony. Instead, he went to a NASCAR race in Daytona. A year and a half later, less than a week after Iacocca announced that he'd raised the $230 million he'd hoped to, he was sacked from the commission.

By then, the restoration foundation was in trouble. In the summer of 1985, Congressman Bruce Vento, chair of the Subcommittee on National Parks and Recreation, held hearings on the relationship between the foundation and the National Park Service. Vento didn't like the authority of the foundation. He spoke on *20/20* of an "improper delegation of power," and said that commercialization of the statue was "not unlike a whore who's being pimped on the sidewalk." He told the *Philadelphia Inquirer* that the foundation seemed "a quasi government structure . . . Who the hell elected them?" There were allegations of threats, reports of the project being "in a total state of chaos." He complained that the foundation was amassing the largest mailing list in history, and that it might be used for political purposes. He asked the US General Accounting Office (GAO), which investigates government expenditures for Congress, to audit the project.

To cap it all off, when the scaffolding was removed, and the Statue of Liberty revealed in all her glory for the first time in two and a half years, a black scar on her face drew attention. It was a streak caused by the baking soda, and only time would turn it green, but a rumor formed that the mark was from workers, who, rather than climb down the scaffolding and use the bathrooms on the ground, had urinated on her face.

⌒

No other rust battle in America has been fought so visibly, contentiously, or been celebrated so grandly. On July 4, 1986, millions of people showed up for the centennial celebration, as did 40,000 boats—including an aircraft carrier, the *Queen Elizabeth II*, and more tall ships than had ever gathered together before. There were so many boats in the harbor that the Staten Island Ferry took twice as long as normal to weave through them. Queens and Staten Island made available 10,000 camping spots. Bleachers were erected. Governors Island became the VIP island, where the Secretary of the Navy sat. On Liberty Island, Walter Cronkite performed as the master of ceremonies, and Nancy Reagan cut a ribbon. Moffitt sat there. Baboian sat there with his wife. Long-haired Ed Drummond was not invited, even though he had forced Moffitt to take a close look at the statue with binoculars, then announced, in bold red letters that needed little poetic interpretation, that liberty was framed. Nobody ever thanked him.

The evening before the big celebration, Cardinal John O'Connor held an ecumenical mass at St. Patrick's Cathedral. That day, Chief Justice Warren Burger swore in 250 new US citizens on Ellis Island. The Boston Pops played in New Jersey. The New York Philharmonic Orchestra played in Central Park. Events were held at the Meadowlands. Sinatra didn't show up. The weekend alone cost just under $40 million. So many visitors flocked to the statue, and were forced to stand in such long lines, that a riot almost ensued. Almost a third of the world's population saw the ceremony on TV.

As Grover Cleveland had shown up on a boat for the dedication a hundred years earlier, that's how President Reagan planned to arrive. He wanted to show up aboard the USS *John F. Kennedy* and "relight" the torch with a laser beam. Instead, he performed the ceremony from Governors Island. In any case, the achievement was no less triumphant than it had been a century before; in both cases, the Statue of Liberty exemplified the greatest fusion of engineering and art of the day. The occasion was followed by the largest fireworks show ever—twenty tons of fireworks, launched from forty barges, by a pyrotechnics partnership called the All-American Fireworks Team.

That may be the best symbol of all: planned oxidation, commemorating defeated oxidation. Which makes you wonder: had people known about the rust battle, would the celebration have been even grander? What's more impressive: liberty or engineering? Philosophy or power? Belief or might? History or science? From the metal's point of view, there was nothing democratic going on. The metal had become part of a totalitarian regime, planned, controlled, observed, denied the opportunity to do what it yearned most to do. It would be a strange thing to celebrate the metal's fate. Better to focus on the fireworks.

Today, a plaque, installed by the National Association of Corrosion Engineers, marks the site. Below the NACE logo—a triangle within a circle, with two fig leaves around it—there's this text:

THE STATUE OF LIBERTY

HAS BEEN SELECTED BY THE

NATIONAL ASSOCIATION OF

CORROSION ENGINEERS

AS A

NATIONAL CORROSION RESTORATION SITE

AS AN EXAMPLE OF MAN'S TECHNOLOGICAL

ACCOMPLISHMENTS TO CONTROL CORROSION

APPLIED TO A HISTORIC STRUCTURE SO

THAT FUTURE GENERATIONS CAN BENEFIT

FROM THE SYMBOLIC HISTORY OF THE

STATUE AS THE WORLD'S BEST-KNOWN

MONUMENT TO MAN'S SEARCH FOR FREEDOM

AND LIBERTY.

PRESENTED TO THE

NATIONAL PARK SERVICE

OCTOBER 28, 1986

IN COMMEMORATION

OF THE STATUE'S 100TH BIRTHDAY

It is a unique plaque, marking the only National Corrosion Restoration Site in the country. But it won't be the last.

2

SPOILED IRON

The first recorded words about rust express what you might expect: exasperation. The words belong to a Roman army general. Two thousand years ago, during the Blue Nile campaign, he complained about corrosion in his giant catapults. "The pentle hooks on the onagers are weakened so badly by corrosion," he wrote, "that the arbalests are causing more casualties in our own army than to the enemy." A generation later, at a loss to explain how rust worked, Pliny the Elder figured, metaphysically, that the benevolence of nature had inflicted the penalty of rust to limit the power of iron, thus making nothing in the world more mortal than that which is most hostile to mortality. He called rust *ferrum corrumpitur*, or "spoiled iron." Only in myths was rust less troublesome. It helped Iphicles father a son, and it healed the stubborn wound in Telephus's thigh. Rust confounded the rest of us.

Robert Boyle, the tall, wealthy English "father of chemistry," took up an investigation of rust during the seventeenth-century reign of King Charles II. He began by insulting Pliny: "I have not found among the Aristotelians," he wrote, "so much as an offer at an intelligible account." Born a year after the death of Francis Bacon, Boyle talked metaphysics with Sir Isaac Newton, hung out with the founders of the scientific group known as the

Royal Society, and was in Florence, Italy, when Galileo died. He taught himself Hebrew, Greek, and Arabic so that he could read original sources. He conducted medical experiments on himself, and tasted his own urine. By the end of his life, he'd written more than two and a half million words. In his 1675 treatise *Experiments and Notes About the Mechanical Origine or Production of Corrosiveness and Corrosibility,* Boyle summed up rust as a mechanical phenomenon resulting from the "congruity between the agent and the patient." For a metal to be corroded, it had to be "furnish'd with pores of such bigness and figure, that the corpuscles of the solvent may enter them." These "pores" also explained why light penetrated glass. (That same year, Boyle also published an account of the "transmutation" of mercury into gold.) By his own intelligible standard, Boyle failed, too, only more discursively than Pliny, and with a higher word count.

His twenty experiments were not for naught, though. He found that salt, by itself, didn't corrode lead nearly as fast as saltwater did. By pouring saltwater, lemon juice, vinegar (which had "edges like blades of swords"), urine, turpentine, lye, and various acids onto lead, iron, mercury, copper, antimony, and tin, he showed that all metals were vulnerable. Silver, for example, fell victim to nitric acid. Even gold corroded when subjected to a mixture of nitric and hydrochloric acids called *aqua regia.*

Today we know that only a handful of rare metals don't corrode: tantalum, niobium, iridium, and osmium. The others—all of them—can be invited, urged, or forced to react with oxygen. Some react spontaneously in air or water. Some, like aluminum, chrome, nickel, and titanium, form a thin outer layer of protective metal oxide, and then call it quits. Many of the corrosion-resistant metals are named in honor of Greek gods or kings, for no other entity could have created such marvelous stuff. Nevertheless, most metals met oxygen long ago, which explains why precious few metals present their naked selves anywhere on Earth. (This also explains why oxygen did not accumulate in the atmosphere for billions of years, until rocks on the surface had reached their fill.) Three-quarters of the universe's elements are metals, and nature, apparently, abhors almost all of them.

A few lucky ancient wanderers found rare but marvelously resilient chunks of metal. Shiny and strong, the stuff was perfect for Inuit spears, or Sumerian shields, or Tibetan jewelry. It was a nickel-iron alloy, not unlike stainless steel, and it came from the sky, in the form of meteorites. The world's largest meteorite, in fact, in Namibia, is of this variety. Today we call it "meteoritic iron," but long before stainless steel was invented, one person called this "celestial metal" an advantage "providentially placed before us." The disadvantage placed providentially before everyone else was rust.

Rust, of course, is the corrosion of iron, while corrosion is the gnawing away, thanks to oxygen, of any metal. To the horror of engineers, I use the word colloquially.

As a result of rust, oxygen commands love and hate. It contributed to the filling of the oceans, got life beyond a green slime, and assisted with the evolution of two sexes, earning one modern biochemist's nickname "the molecule that made the world." Just two centuries earlier, though, on account of oxygen's potency and ubiquity, a less impressed chemist called it "the fire that burns up all things slowly." It does not get along with metals, or rather, the way it gets along with metals, as Boyle demonstrated, does not agree with us.

But oxygen wasn't discovered until a couple of years before the United States declared independence from Britain, and wasn't recognized as the culprit in rust for another fifty years. Until then, experiments relating to rust caused confusion and enlightenment in equal proportions.

In Bologna, Italy, Luigi Galvani figured out that dissimilar metals could be used to create sparks—enough to make the legs of a frog twitch—and rust. Two hundred miles northwest, in Como, a physics professor named Alessandro Volta figured out how to make a battery by stacking sixty different metals in a tower, but didn't know what the metals had to do with it, or why the current eventually ebbed. Using such a battery—a "voltaic pile"—the frantic British chemist Sir Humphry Davy demonstrated a technique named after his Italian colleague. By galvanizing steel—coating it in a thin layer of zinc—he could protect it from rust.

At the turn of the century, though, chemistry was still muddled with alchemy and philosophy. There were about a dozen known elements, and three recognized processes: combustion, respiration, and oxidation. Combustion and oxidation were explained by a mysterious agent called phlogiston. Heat was understood as a weightless fluid called caloric. An experimentalist at heart, Davy didn't like either theory. In 1806, after more experiments with batteries, he coined the term *electrochemistry*. The next year, he discovered sodium and potassium, new metals of the reactive variety. Colleagues compared him to Boyle, and the government saw him as society's savior. He was asked to investigate explosions in mines, and after studying the combustion of methane, designed a miners' safety lamp that, because it was surrounded by iron mesh, dissipated heat and eliminated explosions. It made him famous.

In 1823 the Royal Navy commissioned Davy to solve the problem of corrosion on their warships. Copper sheathing on ships protected wooden hulls from destruction by worms and rot, and prevented the adhesion of barnacles, weeds, and other sea life, which slowed down ships. By then, Davy had been knighted, and had successfully campaigned for the presidency of the Royal Society. He began by putting copper in beakers of seawater and investigating the greenish precipitate. Deducing that it formed with oxygen, he figured the oxygen had to be coming from the water or the air, and since no hydrogen was produced, it couldn't be coming from the water. It was the air. To prove it, he showed that very salty water—brine—didn't dissolve the copper as fast as regular seawater, because it contained less dissolved oxygen. (He'd already shown years before that deoxygenated water killed fish.)

Then, with the principles of galvanizing in mind, he attached nails made of zinc and iron—"more oxidable metals"—to sheets of copper, and put them in the sea. Davy spent months experimenting at Chatham and Portsmouth, studying the ratios of the metals. Finally, in January 1824, he announced his findings. "A piece of zinc as large as a pea, or the point of a small iron nail, were found fully adequate to preserve forty or fifty square meters of copper," he wrote. A zinc-to-copper ratio anywhere between 1:40 and 1:150 prevented corrosion. Between 1:200 and 1:400, corrosion was slowed down, but not prevented. Less than that had little effect. It didn't matter where the zinc was placed, as long as it was connected to the copper. He called his results "beautiful and unequivocal."

The navy tested his method on the HMS *Comet*. Months into the trial, the ship wasn't corroding, but was fouling with barnacles far worse than before. The navy was annoyed, and the Royal Society was embarrassed. Letters in newspapers tarnished Davy's reputation. But Davy had been right. His sacrificial anodes—still widely used on boats today—prevented corrosion.

Michael Faraday, the renowned experimentalist and father of electro-
magnetism, whom Davy had hired as an assistant in 1813, took Davy's
work even further and, in the 1840s, determined that electrical current—
not a fluid at all—could be used advantageously to prevent corrosion. "All
chemical phenomena," he wrote, "are but exhibitions of electrical attrac-
tions."

For much of the nineteenth century, though, chemists focused elsewhere—on the composition of molecules, on using spectroscopy to iden-
tify and isolate and eventually describe new elements. At the end of the
century, most believed that acids, carbonic acid in particular, were responsible for corrosion. (Acids were only part of the story.) Others thought hydrogen peroxide was somehow involved. Some thought that imperfections
in metal were to blame, and that perfectly pure metals wouldn't corrode.

Not until the first half of the twentieth century did corrosion theory
take shape. It started with Swiss chemist Julius Tafel, a handsomely
bearded insomniac, who in 1905 related current and voltage to the rate of
a chemical reaction. He killed himself before chemists Johannes Bronsted,
Martin Lowry, and Gilbert Lewis, in 1923, each came up with the notion
that chemical bonding resulted from the pairing up of acids and bases,
which, depending on how you looked at it, either donated/received pro-
tons, or donated/received electrons. Three years later, Linus Pauling and
Robert Mulliken—both future Nobel Prize winners—began quantifying
the tendencies of elements to attract electrons, a property called electronegativity. Because every element is structured differently with its electrons
orbiting its neutrons, each has a unique electronegativity, though many are
similar. Little known francium, on the lower left corner of the periodic
table, is the least electronegative, while fluorine, on the opposite corner,
is the most electronegative. The scale runs from 0.7 to 4, in Pauling units.
Fluorine, reacting with everything, steals electrons with fury. Oxygen, a
gas five hundred times more abundant on earth than fluorine, is the second most electronegative element—and this explains why life relies on it.
For transporting energy, it is the best thing going. (Aerobic metabolisms,
which rely on oxygen, are fifteen times more efficient than anaerobic metabolisms.) The third most electronegative element is chlorine. Considering that two-thirds of the world is covered by a liquid containing a great
deal of dissolved chlorine, it's almost as if God stacked the cards against

admirals intent on employing metals in their service. God didn't make calm air or seas.

In sort of a mirror of electronegativity, metals can be ranked in nobility. Noble metals don't give up their electrons, no matter how electronegative the other elements may be. The most noble metals—gold, platinum, iridium, palladium, osmium, silver, rhodium, and ruthenium—are also the most valuable, and this is no coincidence. They're valuable because they're reliable. They don't corrode. The nobility of a metal is measured in volts, from 1.18 (platinum) to -1.6 (magnesium).

Therein lay the source of the current in Volta's piles of metals: the unnoble metals were giving up electrons to the noble ones. A quarter-volt difference is enough to compel an electron to migrate. There's a quarter volt between lead and titanium, and there's a quarter volt between tin and silver. The quarter volt between iron and copper is what saved the copper on the bottom of the HMS *Comet*, and what saved the Statue of Liberty's skin at the expense of its frame. The quarter volt between aluminum and steel is what causes seat-posts to rust firmly into place on bicycles, and what caused the white powder around the rivets in *Syzygy*'s mast.

In other words, the unnoble metals are anodic. Paired with more cathodic (in other words, more noble) metals, they sacrifice themselves at the altar of physics. Evans, of Wimbledon, diagrammed this back in Pauling's day. That's why galvanizing works: you're giving nature something to chew on for a while.

Oxygen wants to steal electrons, and so get reduced. Unnoble metals give them up, and so get oxidized, especially when compelled by more noble metals. Water's the convenient pathway for the particles and elements to zip around in. In 1938 Carl Wagner and Wilhelm Traud described this phenomenon, putting together Tafel's observation. They said that the sum of the charges lost and gained in a corrosion reaction was zero, and that each metal would corrode at a new rate. It became known as the mixed-potential theory of galvanic corrosion.

By then chemists also recognized that most of one volt was enough to compel electrons to stay put. In other words, if a pipeline operator pushed 0.85 volts into his buried pipeline, he could convince the electrons in the steel not to be lured elsewhere.

Together, this cathodic protection and anodic protection form half of the techniques available to combatting corrosion. The third arm of defense, far blunter, precedes them both. It's paint. If you can stop oxygen (and water, which contains oxygen) from getting to metal, you can stop corrosion. The fourth arm is sort of a modern version of paint. Inhibitors, binding to metal before oxygen has a chance to, work just as well in abetting a brown outcome. Many are synthetic, but they've been made from mangos, Egyptian honey, and Kentucky tobacco. Anodizing—intentionally oxidizing the surface of aluminum by dipping it in acid and applying current—works because the thick oxide layer is then sealed with an inhibitor. Electroplating with a metal more durable than zinc—cadmium, chromium, nickel, or gold—is sort of the rich-man's galvanizing.

Of course, subtleties in the theory of corrosion abound. Poorly mixed alloys may become anodic and cathodic to each other in opposite corners of the same piece. Once electrons begin to flow, the disparity only increases, and corrosion accelerates. Metallurgists using X-ray crystallography nearly a hundred years ago figured that out.

In the same year that Wagner and Traud described galvanic corrosion, Marcel Pourbaix, of Brussels, Belgium, came up with thermodynamic diagrams of corrosion reactions. He broke down the oxidation and reduction reactions in a metal across the full range of electrochemical conditions, from super acidic to super basic. The resulting graph showed zones of corrosion, passivity, stability. It showed where a metal was safe and where it was under threat—revealing why the acids that Boyle toyed with were bad news for metals. Pourbaix did this for each element, publishing his results in his *Atlas of Electrochemical Equilibria in Aqueous Solutions* in French in 1963 and in English three years later. Then this pioneer of the field traveled around the world talking about the new science of rust.

Yet much in the field remained enigmatic. Francis LaQue, another pioneer in the science of corrosion, said before his death in 1988, "Corrosion engineers, like economists, know enough to provide plausible explanations of what has happened without being equally adept at predicting future occurrences." That almost sounded like Pliny.

3

KNIVES THAT WON'T CUT

Sometime in 1882, a skinny, dark-haired, eleven-year-old boy named Harry Brearley entered a steelworks for the first time. A shy kid—he was scared of the dark, and a picky eater—he was also curious, and the industrial revolution in Sheffield, England, offered much in the way of amusements. He enjoyed wandering around town—he later called himself a Sheffield Street Arab—watching road builders, bricklayers, painters, coal deliverers, butchers, and grinders. He was drawn especially to workshops; if he couldn't see in a shop window, he would knock on the door and offer to run an errand for the privilege of watching whatever work was going on inside. Factories were even more appealing, and he had learned to gain access by delivering, or pretending to deliver, lunch or dinner to an employee. Once inside, he must have reveled, for not until the day's end did he emerge, all grimy and gray but for his blue eyes. Inside the steelworks, the action compelled him so much that he spent hours sitting inconspicuously on great piles of coal, breathing through his mouth, watching brawny men shoveling fuel into furnaces, hammering white-hot ingots of iron. He was mesmerized. Day after day, he watched men forge, cast, ground, buff, and burnish metal until it was shiny and bright. Sparks were flying, probably in Harry's mind, too.

There was one operation in particular that young Harry liked: a toughness test performed by the blacksmith. After melting and pouring a molten mixture from a crucible, the blacksmith would cast a bar or two of that alloy, and after it cooled, he would cut notches in the ends of those bars. Then he'd put the bars in a vice, and hammer away at them. The effort required to break the metal bars, as interpreted through the blacksmith's muscles, could vary by an order of magnitude, but the result of the test was expressed qualitatively. The metal was pronounced on the spot either rotten or darned good stuff. The latter was simply called D.G.S. The aim of the men at that steelworks, and every other, was to produce D.G.S., and Harry took that to heart.

In this way, young Harry became familiar with steelmaking long before he formally taught himself as much as there was to know about the practice. It was the beginning of a life devoted to steel, without the distractions of hobbies, vacations, or church. It was the origin of a career in which Brearley wrote eight books on metals, five of which contain the word *steel* in the title; in which he could argue about steelmaking—but not politics—all night; and in which the love and devotion he bestowed upon inanimate metals exceeded that which he bestowed upon his parents or wife or son. Steel was Harry's true love.

By the time he retired, Harry Brearley knew intimately every facet of steelmaking. Having earned a reputation, he—like Frank Sinatra begrudging the popularity of the Beatles—grew defensive, and then reactive, as technology evolved. The emotion got to him. Near the end of his days, he yearned to have been a generalist, and wrote that if he could do it all over again he'd be a doctor, and none too specialized, so that he could see all the colors of life. "There is nothing deader than a list of the chemical elements," he wrote, "whose vitalised dances are supposed to make life."

The man who discovered stainless steel was also a rebel, and fittingly so, for what he found was contrary to nature. The runt of a large family, largely ignored and taught chores so as not to be a burden, Harry Brearley became the successful one—the one whose name lives on. He was a chemist who never attended a single course in chemistry; yet he scorned official titles. And rather than consider himself a chemist, analyst, or research director, he preferred to think of himself as a "competent observer," a "professional observer," and an "experimentalist." Self-taught, he refused to subject his own son to what he called the mashed-potato educational system. He abandoned church, except to flirt with his would-be wife. A product of the gutter, he rose from the factory to the boardroom, and once there, stepped down so that he could go back to the ground floor of the factory.

After a long career as a scientist, he insisted that he was an artist, because he thought about steel with his heart rather than his head. Questioning chemists' test results, he called their reports "bogey tales" of "bluff and bunkum." He resisted modernization. He called himself "a breaker of idols and a scorner of cherished regulations." A binary man, he had no room for gray tones. He was curious but opinionated, flexible but intolerant, innovative but persnickety, knowledgeable but overconfident, and determined but obstinate. He was patient with metals and impatient with masters. He even became a class warrior—a lover of underdogs like himself—and then somewhat paranoid. All because of steel.

Harry Brearley's rebelliousness was almost his undoing. Despite advances in the industry, he frowned upon modern, high-volume steelmaking technologies as much as he disparaged old beliefs that persisted without regard to logic. He knew both were wrong; he felt it. His soul rooted for man over machine, flexibility and ingenuity over rigid process and operation, skill and judgment over precision. Any other framework made him queasy. He resented that metallography ("taught by professors and text-books") was becoming more valued than metallurgy ("practiced in the works"). He understood that the best steelmaker could know nothing about chemistry, and vice versa, and he hated that the analyst was revered, while the steelworker was underappreciated. Steelmaking, of all things, he figured, should be meritorious. His knowledge commanded respect, but business-wise, he was naive. An anachronism at work, he became, to his bosses, almost a liability. Yet he was proud that he'd had "the courage to ignore time-worn precept and reach success by roads which . . . should lead to direct failure."

The road almost led to failure, for the discovery of stainless steel was not obvious; many others had missed it, and Brearley almost overlooked it as well. Success, too, was not immediate; and Harry Brearley's name, which had been respected if not esteemed, all the while suffered. Before the commercial success of stainless steel, Harry Brearley was briefly known as "the inventor of knives that won't cut." His doggedness in pursuing that commercial success cost him his job, killed a business relationship that had lasted thirty years, and compelled others to try to appropriate the credit due him. He was almost outmaneuvered.

Patenting his discovery took more effort than he thought it would, and commercializing it brought further difficulties. He later wrote that the "malodorous happenings" involved "used up parts of my life that might have been used more enjoyably." On the other hand, the discovery—or at least the popularization—of stainless steel made him rich, earned him one of the highest awards in metallurgy, and etched his name into history. He was a devoted, studious, and attentive artisan, and even if he wasn't the first to create or discover or patent or commercialize stainless steel, he deserves the reward for it, because his persistence was more than commensurate with the trouble it caused him.

~~~~~~

Harry Brearley was born on February 18, 1871, and grew up poor, in a small, cramped house on Marcus Street, in Ramsden's Yard, on a hill in Sheffield. The city was the world capital of steelmaking; by 1850 Sheffield steelmakers produced half of all the steel in Europe, and 90 percent of the steel in England. By 1860, no fewer than 178 edge tool and saw makers were registered in Sheffield. In the first half of the nineteenth century, as Sheffield rose to prominence, the population of the city grew fivefold, and its filth grew proportionally. A saying at the time, that "where there's muck there's money," legitimized the grime, reek, and dust of industrial Sheffield, but Harry recognized later that it was a misfortune to be from there, for nobody had much ambition.

The men there were all laborers—joiners or wheelwrights or blacksmiths—and they'd all come to the city enticed by higher wages. The women worked hard to support their families. All appeared worn down, beset with the aches of manual labor and respiratory problems like "grinders disease," the result of inhaling sandstone and steel particles all day. Harry's mother, Jane, an unrefined but direct woman with quick, brown eyes and a firm mouth, was the youngest daughter of a blacksmith. She'd had six months of schooling, and never learned math. She could read, write, and reckon. She wasted nothing. His father, John, a tall, strong, and bulky man with curly brown hair and blue eyes, was a steelmaker. At times dreamy, he offered a hint of a poetic streak, but he was by no means intellectual. He was also a drinker, and short of work.

There were eight houses in Ramsden's Yard, four on either side of a hard-packed square black as night with soot. The doors of most of the houses were always open, full of restless children, of which there were many. The Andrews family had four kids, the Whiteheads had three, the Linleys had five, the Brayshaws had five, and the Brearleys had nine, including Harry, the youngest. There was also an old lady, and a young woman, but nobody ever saw them. Like their fathers, many of the boys were bow-legged and stooped, from sitting in factories, cutting file blanks all day.

A mother's boy, Harry was closest to his brother Arthur, who was radiant, strong, and stubborn where Harry was weak, wayward, and feeble. He was so frail that he stayed home from school often, and learned, from his mother, many domestic chores: to sew, darn, scrub, wash clothes, and shop at the market.

The house Harry grew up in was sparse and tight; the living room measured ten feet square, with two bedrooms above it. The kids ate standing up because there were not enough chairs. There were no books, or pictures, or toys; there was no space for a desk. The Brearleys were heartily poor but not starving, yet they weren't far from the breadline. Harry wore jackets that had been made from his father's trousers. He helped deliver coal in a wheelbarrow in return for sweets. After school, he bundled sticks, earning a penny for a dozen bundles. He used to walk along nearby railroad tracks, collect lumps of coal that had fallen from passing trains, and bring them home to his mother. He once borrowed a book from the library, and copied it—the whole thing—by hand, because he couldn't afford to buy a copy.

In 1882 his parents moved down to Carlisle Street, beside the railroad tracks—a place said to be separated from hell by only a sheet of tissue paper. It was filthier, dustier, smokier. But Harry loved it, on account of the increased color and variety. There was a pigsty and stables to poke around in, and more adult conversation to pick up. On account of his curiosity, he was regularly late for school; he found too much to look at on the way. For punishment he was caned, cuffed with a wet handbag, kicked with a clog-toe, and kept inside. His schooling, too, was minimal. Well into his teens, he didn't know who Shakespeare was. He once asked a colleague: "Is he an

Englishman?" But he got away from school, at age eleven, with his "brains unshackled and his curiosity undimmed," and was then free to work, according to the law, in nonfactory conditions.

He was unhappy in his first jobs. He spent three days in Marsland's Clog Shop, blacking boots and carrying things from eight in the morning until eleven o'clock at night, and hated it. He spent a week in Moorwood's Iron Foundry, painting black varnish onto kitchen stoves, before being discharged on account of labor regulations. He spent six weeks helping a doctor, but was disheartened by the subservience the man required. Finally his father took him to work in the Thomas Firth & Sons steelmaking factory, where he worked as a nipper, or cellar boy, moving clay stands and covers wherever needed in the dark, hot ashes of the cellar, and skimming the slag from the steel. Everybody, including his father, thought he was too small and weak for the job, but he spent three months at Firth's, working long, sweaty days, before he was once again discharged on account of violating labor regulations.

He was then hired as a bottle washer by James Taylor, the chief chemist in a laboratory of the same steelmakers. The accomplished son of two weavers, Taylor had grown up poor and won a scholarship, attended the Royal School of Mines, worked for a professor at Owens College, studied with the German chemist Robert Bunsen in Heidelberg, and worked in Bolivia and Serbia. He was thirty-five years old, pale, and had a scraggly beard. Harry hadn't ever heard the word *laboratory* before, and when he first showed up, was so overwhelmed by the amount of glassware that he figured it was a place people came to drink. At first, he found the work tedious, but his mother encouraged him to stay there, as it was undoubtedly better than the melting furnaces in the steelworks. Harry was only twelve; he would become Taylor's protégé.

Taylor started his training by teaching Harry arithmetic (Harry had to buy the book himself) and then, a couple of years later, algebra (Taylor bought him the book, a gift Harry brought home to show off, and never forgot). Taylor bought Harry a set of drawing instruments too. Taylor was not social, not a drinker, not a smoker, not a swearer. He didn't even speak in the Sheffield dialect. But he was thrifty and handy, and the set of skills he displayed was formative for Harry. Under Taylor, Harry learned to join wood, paint, solder, plumb, blow glass, bind books, and work with metal.

While his friends were out playing, Harry was learning new skills. (He'd dislocated a knee playing soccer at age fourteen and then steered clear of most athletic pursuits. He was not a good fisherman, and was a terrible shot. He remained amateurish in every hobby. Yet he was not clumsy, even if he did break half the flasks and beakers he grabbed in the laboratory.) This knowledge later inspired Harry to make his own furniture, stitch his own sandals, and try writing. His first attempt, an article for *Windsor* magazine, described the nature of various inks in creating inkblots, of which he made a few hundred; the next was titled "Bubble-Blowing as a Physical Exercise." Some hobbies. He also attended night school, on Taylor's urging, studying math and physics a few nights a week.

By the time he was twenty, he was proficient in most crafts, even though, technically, he was a bottle washer. The lab suited him; at work, one assistant sometimes sang opera or recited poems, while Taylor regularly discussed food, economics, education, politics, and social welfare. In this context, Harry grew comfortable in the presence of educated people. Despite, or perhaps because of, Harry's reverence for his boss—which bordered on idolatry—Harry's mother by then was encouraging him to find a job in a factory, one with a better salary, and more of a future.

His mother died the next year. Brearley moved in with his older brother Arthur. That same year, Taylor left for work in Australia, and Brearley was promoted to lab assistant. Contemplating life ahead, he had a sudden conversion, and decided more schooling was not for him. He recognized that he had no tolerance for things he didn't want to do. He was hardening, like steel.

He also fell in love, and began courting his future wife, Helen, chatting her up at Sunday school. (Though, in recollecting the time, he cites his first love as analytical chemistry.) At age twenty-four, they married; he'd been promoted to an analytical chemist at the lab, and was earning two pounds a week. Together they had a total savings of five pounds. They lived on bread, onions, and apple pie, in a simple cottage south of Sheffield, but he never mentions it, or his wife, in his autobiography. He barely mentions his only son, Leo Taylor Brearley (named after James Taylor), who was born two years later. But he mentions love: "I was in love with my work, and

could think of few better things than the privilege of living to continue it."
He enjoyed it so much that he said it made him feel drunk.

So he drank: he spent the next six years reading everything he could
about metallurgy, starting with periodicals and journals about chemistry,
barely stopping for a lunch of bread and dates. Next he read about manganese, and every process by which it could be detected in steel. Then he
read about every other steelmaking element; all the while he kept index
cards detailing what he had learned from each book. He developed his
knowledge carefully, procedurally, accumulating as much as he could. Lab
protocol stipulated that anyone who figured out how to save time could
enjoy his savings as he wanted. Brearley got his day's work done in a couple
of hours, and spent the rest of the day reading and experimenting.

In his late twenties, Brearley started writing technical papers on the
analytical chemistry of metals for publications such as *Chemical News*. Taylor wrote from Australia, offering him a job assaying gold and silver. He
turned it down. He was developing a reputation as a steel problem solver,
and enjoying it.

On Saturdays, for fun, he met up with Fred Ibbotson, a professor of
metallurgical chemistry. Ibbotson would give him samples of metals, and
challenge him to determine, in ten, twenty, or thirty minutes, how much of
a given element they contained. What happened to blowing bubbles? On
Sundays, he hung out at the lab with his brother Arthur (who'd walked
three miles to get there), and together they analyzed enough steel to get
proficient at it. This was the beginning of a lifelong working relationship with his brother. Twenty years later, the two cowrote *Ingots and Ingot
Moulds*; Brearley thought it was his best work. Two years after that, when
he won the Bessemer Gold Medal, the highest award conferred by the Iron
and Steel Institute, for outstanding contributions to the steel industry, he
credited his brother generously. His sixth book, *Steel-Makers and Knotted
String*, was dedicated to Arthur, as "playmate schoolmate and workmate."
In it, he called his brother "a better workman, a better observer and a more
resourceful experimentalist than I."

In 1901, at age thirty, Brearley was hired at Kayser, Ellison & Co. as
a chemist to work on high-speed tool steels, which had been discovered
three years earlier by a consultant for Bethlehem Steel named Frederick
Winslow Taylor. Sidetracked from production problems, Taylor had begun

looking at steels used to plane and bore ship plates and cannons. Ideal forging temperatures were still measured by color, and he found that steel, heated to just below dull cherry, came out strong, but the same steel, heated above that point, became weak. To his surprise, he found that if he heated it further—to salmon and yellow—the steel got superhard; so hard that machinists could run their cutting tools two or three times as fast as before, until the blades glowed red, at 1,000 degrees Celsius. It was so dramatic that at the Paris Exhibition of 1900, Taylor set up a giant lathe in the dark, so that the glowing-red cutting edge, as well as the stream of blue chips, was visible.

In 1902 Brearley cowrote his first book with Professor Ibbotson. It was called *The Analysis of Steel-Works Materials*. That same year, he teamed up with his old operatic lab mate, Colin Moorwood, and started a company, the Amalgams Co. He'd developed a unique claylike material, and they profited selling it to a local business. He and Moorwood spent every evening and weekend toying with new materials, and made a mess of one room in his house. Within a year, he'd written his second book, *The Analytical Chemistry of Uranium*.

Steel business was good, and in September 1903, Brearley's old employer, Thomas Firth & Sons, bought a steelmaking plant in Riga—Russia's second largest port, on the Baltic Sea—in order to produce steel for the massive Russian market without having to pay export tariffs. On Moorwood's recommendation, Brearley was hired to be the chemist. Arthur Brearley would join them too, probably on his brother's recommendation. Moorwood would be the general manager. Together Brearley and Moorwood traveled there in January 1904, in the dead of winter.

The Salamander Works, as the factory was called, was a bright, roomy place, covering forty acres on the southern bank of the Jugla River, between two lakes. It was six miles northeast of town and not well equipped. There was no gas or water; the latter had to be brought inside in buckets. It was so cold that Brearley wore an overcoat and tall rubber boots all day and took a portable paraffin stove to wherever he was working.

Worse, there were no experienced workers, at least none who knew how to properly forge, anneal, machine, and harden steel. With the Russo-Japanese War under way, the Salamander Works had been contracted to make armor-piercing shells for the Russian navy. Firth sent an Englishman named Bowness to oversee the process. He turned out to be incompetent. At firing tests, in Saint Petersburg, the shells he'd produced failed miserably. According to Brearley, the "expert hardener" explained his masterly technique thus: the secret in hardening, Bowness said, was "to heat the buggers." Brearley was put in charge, promoted to heat treater.

Brearley and his brother set about determining the temperature range at which the shell steel could be hardened without ruining it. The sweet temperature spot could be determined by examining hardened steel for smooth, fine fractures. A problem arose: the works had no high-temperature pyrometers, or quick source of them. Brearley figured he and his brother could eyeball it, but he also realized the two of them couldn't possibly be present for all the work that lay ahead.

So they improvised. Brearley mixed together combinations of a wide variety of chemicals and metals, including coins, and created three salt-based alloys that melted in the desired temperature range. He melted a collection of these, then cast them into small cylinders and cones about the size of a little toe, and coated them with brown, green, and blue waxes. He called these sentinel pyrometers, because by balancing them on porcelain dishes inside the furnace, a worker could easily see if one was melting or not. If the first, brown sentinel melted, the furnace was at the temperature just high enough to harden the steel; if the green sentinel melted, too, the temperature was right on; and if the blue sentinel melted, they'd know that they'd overshot their mark and that the steel would be ruined. Using the sentinels as guides, Brearley produced a second batch of hardened shells, and they passed the firing test, as did every batch made after that—even those made by his novice metal workers.

His style was casual. Men swapped roles, worked as a team, and were free to divulge their opinions. There was no organizational hierarchy. No mechanical precision. No engineered plan drawn up on paper. Moorwood okayed the arrangement, and agreed not to interfere. Under such management, Brearley found that he preferred novice steelmakers. They weren't biased by previous experience, or hamstrung by any preconceived notions. In time, he credited the Latvian peasants with skills rivaling those of his Sheffield pals.

He sent the formulas for the sentinel pyrometers back home, and Amalgams Co. sold thousands. Within the year, he was promoted to technical director, and put in charge of building a crucible furnace. He ordered some plans. The plans were wrong, but his furnace was right. He was also put in charge of selling high-speed tool steel. He surprised many customers by stripping from his business attire and working in the furnace, just like any of the other men, to demonstrate his product. Who was this Brearley: a technical director or a technician?

With his long, boyish face, and big, dark, owlish eyes, Brearley still looked like a teenager. He was clean shaven, with short black hair parted down the middle. He wore wire-frame spectacles. His ears were not lacking in prominence. By now, his adult persona had emerged: he was deliberate and devoted; confident but not dictatorial, and definitely not greedy. He was earning plenty of money, but he remained thrifty, never yearning for a big house or fancy cars or fine food. He was certainly no public speaker, and had no stomach for politics. He wasn't much of a salesman, as he had no ornamental graces, and few cultural graces. In fact, he possessed few social skills: invited to a masked ball, he was advised to let himself go in order to enjoy the occasion. He had no idea how to do that. At another party, he stood aloof, a wallflower. He was incapable of flirting. But he was good at his work.

The revolution came in 1905. The political and cultural revolt didn't bother Brearley so much; in fact, he wasn't especially repelled by socialism. (He'd joined the International Labor Party in England.) But the strikes made it impossible to produce steel, and this bothered him. The furnaces had to run constantly, or not at all. He couldn't start and stop them as the vagaries of politics demanded.

An impromptu public meeting was held on a vacant floor of the factory. Two thousand men showed up. Before the meeting started, revolver cartridges were distributed. Not long after, the foreman blacksmith was murdered outside his apartment. A half dozen factory workers were arrested and imprisoned. The state of affairs terrified many; three engineers fled the country. So did Moorwood. Brearley took his spot as general manager and kept it for three years. He sat in Moorwood's chair, at Moorwood's big horseshoe-shaped desk, in Moorwood's clothes, smoking Moorwood's cigars. It was the most extravagant thing he ever did.

With Brearley in charge, new equipment was in order. He bought a
microscope, a galvanometer, and a thermocouple, and spent weeks toying
with the latter instrument in a cellar. The cellar became the Friday-evening
meeting-place for people who wanted to talk about steel rather than the
revolution. The meetings sometimes went on clear through the night, ad-
journing when it was time to work the next morning. During a strike, with
nothing better to do, the cellar crew made a temperature recorder out of
an old clock and a biscuit tin. They collected pieces of steel that had been
hardened at different temperatures, fractured them, and compared them.
They savored the mysterious specimens. They sought bewilderment for the
sake of discussion. They argued into the night.

Cut off from England and its supplies during the long winters, they
were forced to adapt, innovate, or use substitute materials. In this way, they
gained experience, and what remained in Brearley of any old steelmaking
dogma faded away.

⁓

When Brearley returned to England in 1907, he was offered a position
running the Brown-Firth Research Laboratories, a new joint operation run
by John Brown & Company, which built battleships, and Firth's, which was
working on armor plates. Notably, as research director, Brearley was given
great freedom; he and his employer agreed, before he took the job, that he
could turn down any project that didn't interest him. More importantly,
on account of Brearley's interest in Amalgam Co., they also agreed to split
ownership of rights to any discoveries.

The research was not all excitement; there was plenty of donkey work.
But he also fixed what others thought was unfixable. He found a pile of
train wheels that had been rejected and tossed into a scrap heap, and hardened these as they had learned to do in Riga. The resulting wheels, the ex-
rejects, were better than all of those that had been approved by the railway
company.

Yet Brearley was troubled by new changes in steelmaking. Science was
replacing art. He thought modern metallography, with its focus on the
minute compositions of sulfur and phosphorous, was hype, and mislead-
ing. "The chemical clauses," he later wrote, "do not ensure the quality of
the finished article, any more than the list of the ingredients ensures the

quality of a kitchen dish." He elaborated: "What a man sees through the microscope is more of less," he wrote, "and his vision has been known to be thereby so limited that he misses what he is looking for, which has been apparent at the first glance to the man whose eye is informed to the experience." Theory was gaining traction over experience, and Brearley began to wax nostalgic for the old days, when there were men who could fracture an ingot and tell you, within 0.03 percent, its composition. He saw amateurs making bad predictions, when, as he knew, predictions were worthless. "The man who sets himself up as a metallurgical Solomon," he wrote, "has great odds against him." Most troubling was the advent of new steelmaking technology. Brearley was an old-fashioned steelmaker—maybe the best of them—and the company he worked for wasn't an old-fashioned steelmaking company anymore.

Since 1742, when Benjamin Huntsman devised the crucible process, Sheffield steelmakers had been making steel the same careful way. They melted bar iron in a clay pot, over a coke furnace, and poured it into ingots and molds.

Until then, the only method of making steel was crude, slow, and expensive. Called cementation, it entailed baking bars of Swedish wrought iron in a stone pit full of charcoal until it absorbed enough carbon. (Steel is iron with a carbon content from 0.1 percent to 2 percent.) It took a long time: a few days to get up to temperature, another week of firing, and a few more days of cooling. It took three tons of coke to make one ton of steel. Steel made this way was called blister steel, because carbon deposits on the outside often looked like blisters. A slight improvement could be had by forging many layers of this steel together, to get shear steel, or double shear steel, but that took even more time and labor.

Compared with cementation, the crucible process was a breakthrough. It was careful, precise, and produced steel of a uniform quality. But it was too slow, small scale, expensive, and labor intensive to last. The melters, pullers-out, cokers, pot makers, converters, and nippers were bound to vanish.

The vanishing began in 1855, with the invention of the Bessemer process. By injecting cold air into a chamber of molten iron—a chamber that looked like a big black egg, or maybe a huge grenade—steelmakers were able to burn off carbon and most other contaminants in a white-hot reaction. Then they added some carbon, and voilà: they'd done in twenty minutes what had once taken a week, using one-sixth of the fuel. And they could make fifteen tons at a time, instead of seventy-five pounds. To steel-making companies, it was like being able to sell in barrels instead of pints.

The only problem with the Bessemer process was that iron ore rich in phosphorous—as most was—resulted in brittle, granular junk. It came out rotten, as the blacksmith would say. It was twenty years before a young Welsh chemist named Sidney Gilchrist Thomas figured out a process—known as the Basic process—to precipitate acid phosphorous. Three-quarters of the steel made on England's northeast coast in 1883 was made via the Bessemer process; by 1907, the Basic process had almost replaced it. By then, Carl Wilhelm Siemens and Pierre Emile Martin figured out how to recycle waste-gases to superheat iron, in a regenerative, or open-hearth, furnace. It was a little slower than the Bessemer process, but the steel produced had fine-grained structure, the result of slower cooling, and was much more durable.

Charcoal, too, was on the way out. The gas furnace, invented in 1880s, was the first threat; the electric furnace, invented about 1900, sealed charcoal's fate. The new furnaces caught on in the United States right away; not so in Sheffield. The city was reluctant to modernize, even though new electric furnaces cut down on impurities from burning, made temperature control much easier, and allowed steelmakers to start and stop firing whenever they wanted. (The revolutionaries in Riga would have approved.)

In 1916 more than half the steel in the United States was made via electric furnaces; the next year it was 66 percent; by 1930, more than 99.5 percent of the steel in the United States was made in electric furnaces. In England, it was almost the opposite: the first electric furnace was not used until 1910, and the technology caught on slowly, before it regressed. England produced less steel by electric furnace in 1930 than it did in 1917.

By the end of the nineteenth century, only 1 percent or 2 percent of all steel in America and England was made via the crucible process—but that was no small amount. England exported more than £100,000 worth of crucible steel each month. It tended to be tools and machinery, with high-quality edges, and had a strong reputation even in America. Yet while England's steelmakers may have been reluctant to change, those at Firth were not. Firth began using a gas furnace in 1908 and obtained an electric furnace in 1911. In 1916 the company got seven more—to keep up with demand for munitions and armor needed for the Great War.

Brearley would soon be a dinosaur. But, as a quasi-free-agent analyst at Firth's, his knowledge surpassed that of many other analysts. Other steelmakers described good steel as having "body," attributing it to the type of clay in the crucible, or the source of the water, or the mine from which the ore came. Good steel was therefore mysterious, requiring interpretation. (One Sheffield crucible steel recipe called for the juice of four white onions.) When one steelworker, Henry Seebohm, suggested introducing colored labels to denote the carbon content of steels, Sheffield steelmakers objected. It was too scientific; it eliminated them as translators of intrigue.

John Percy, the author of the 934-page treatise *Metallurgy: The Art of Extracting Metals from Their Ores, and Adapting Them to Various Purposes of Manufacture*, summarized the situation: "the science of the art of steelmaking is still in a very imperfect state, however advanced the art may be." That was in 1864. That same year, by examining the structure of a polished piece of metal with a four-hundred power microscope, Henry Clifton Sorby introduced metallography. Twenty years later, the Sheffield Technical School began offering formal training in metallurgy. Not much had changed fifty years later, except where Brearley was employed.

⌒

Brearley knew qualitative descriptions were bogus misapprehensions, leftover ignorance from an age when science offered little insight. He staked out his turf, relying on skill and science—but not to the exclusion of experience. He ordered two of the earliest Izod notched-bar impact testing machines, each of which, with a calibrated pendulum, quantified the blacksmith's biceps. (The machines are still used today.) He didn't talk about body. He talked about *Krupp-Kanheit*, the result of cooling a nickel-chromium alloy too slowly, leaving it liable to fracture with a brittle, crystalline face.

Brearley saw himself as steel's savior, its priest. He valued depth over breadth. He examined details, concerning himself with quality. But he missed the big picture, and at Firth, had the wrong priorities. Firth cared about volume. Scale. Margin. Market.

Brearley knew that, as far as physical properties of steel go, there's no difference between an axle with 0.035 percent sulfur and one with 0.05 percent sulfur. But he missed the point: the difference, a manager told him, was £2 a ton. It was a lesson in politics as much as commerce; it didn't matter if the steel was no better. It only mattered that people thought it was better, and were willing to pay more for it.

But the lesson didn't stick; if anything, the business of modern steel-making only hardened his resolve that it was all hogwash. "Time was," he lamented later, "when a man made steel, decided what it was good for and told the customer how to make the best of it. Then, with time's quickening step, he just made the steel; he engaged another man, who knew nothing about steelmaking, to analyse it, and say what it was good for. Then he engaged a second man, who knew all about hardening and tempering steel; then a third man who could neither make steel, nor analyse it, nor harden and temper it—but this last tested it, put his OK mark on it and passed it into service." It was a disgrace.

To Brearley, progress seemed like regress. Nobody cared about D.G.S. anymore. He felt like he was the only steelmaker left with his head screwed on right. His expertise was careful and deliberate, untainted. His index cards didn't lie. In 1911 he wrote *The Heat Treatment of Tool Steel*, his third book—and dedicated it to his employer: "To Thos. Firth & Sons, Limited, in whose service labour and learning have been agreeably combined, from 1883 to the present time, these pages are respectfully dedicated by the author." In later editions, this dedication was deleted—a sign of the acrimony that was to come.

⌒

In May 1912 Brearley traveled 130 miles south, to the Royal Small Arms Factory in Enfield to study the erosion of rifle barrels. He examined the problem, then wrote, on June 4, "It might be advisable to start a few erosion trials with varying low-carbon high-chromium steels at once . . ." He spent most of the next year making crucible steels with chromium from 6 percent to 15 percent, but they didn't stack up. Then, on August 13, 1913, he tried the electric furnace, probably grudgingly. The first cast was no good. The second cast (number 1008), on August 20, turned out better. It was 12.8 percent chromium, 0.24 percent carbon, 0.44 percent manganese, and 0.2 percent silicon. He made a three-inch square ingot and then rolled it into a one-and-a-half-inch-diameter bar. It rolled easily and machined well. From that, he made twelve gun barrels, which he sent to the factory.

The factory didn't like them.

Brearley noticed that cut samples of the metal he'd sent had funny properties. He later recalled that, suddenly remembering a date to the theater with his wife, he left some samples in water overnight, and found them unstained the next morning. He studied the metal by polishing it, then etching it with a solution of nitric acid dissolved in alcohol, and looking at it under a microscope. It wouldn't etch, or, rather, it etched very very slowly. It reacted to vinegar and lemon juice the same way. He compared a polished sample of carbon steel to a polished sample of the chromium steel, and was amazed to find after twelve days that while the former had rusted, the latter remained shiny and bright.

Brearley wrote a report, and gave it to his boss. The new metal didn't excite anybody, as far as ordnance was concerned. Brearley couldn't let it go. He wrote another report for Brown's, highlighting the noncorrosive nature of the metal. Ditto for Firth's. He suggested that the metal might be advantageously used in cutlery, which, at the time, was made of carbon steel or sterling silver. (The former rusted, as he well knew; the latter was expensive and still tarnished, which really means that the copper, constituting 8 percent of the alloy, corroded.) The response was not even lukewarm.

He didn't let it go. By the end of 1913, he couldn't stop talking about the utility of the new metal for cutlery. That he thought of cutlery first isn't surprising. He'd spent enough time as a kid helping his mother with domestic chores that he knew the toil associated with cleaning and drying forks, knives, and spoons. Sheffield had also been the center of the cutlery industry since the sixteenth century. He sent samples off to two Sheffield cutlers, George Ibberson and James Dixon. A few months later, a report came back: the steel wouldn't forge, grind, harden, or polish—and wouldn't stay sharp. It was useless for cutlery. Ibberson wrote back: "In our opinion this steel is unsuited for Cutlery steel." The cutlers called him "the inventor of knives that won't cut."

Still, Brearley wouldn't drop it; he said the cutlers were wrong, and said so impolitely. He suggested to his bosses that they sell heat-treated knife blanks. They said no. He suggested a patent. They said no. He continued to make a nuisance of himself.

It's difficult to imagine now, but rustless steel must have seemed an oxymoron of the highest order, like unshatterable glass or unrottable wood or an unsinkable ship or an unkillable person. Iron and steel rust. It's what they do. It's how they are. Everyone grew up recognizing as much. As Brearley wrote later, "The rusting of iron and steel is accepted, like the force of gravity, without question; it is the one property of iron universally recognised. People who have no notion of its tensile strength, or its atomic weight, know that it rusts."

In June 1914 Brearley met a cutlery manager named Ernest Stuart, of the cutlers Robert F. Mosley, whose persistence rivaled his own. Brearley and Stuart had gone to school under the same headmaster. Stuart doubted that a rustless steel existed but recognized that such a thing would be worth bothering about. He bothered by testing a piece in vinegar, after which he reportedly said, "This steel stains less." Stuart was the one who first called it stainless. He took a small sample. A week later, he returned with some cheese knives. He declared them rustless and stainless. But the steel was too hard, and had dulled all of his sharpening tools. He swore. He tried again, and the knives came back very hard, but very brittle. On the third try, Brearley was invited along to watch, even though he knew nothing about knife making. But he knew the temperature at which the steel hardened, and he helped make a dozen knives.

On October 2, 1914, Brearley wrote another report for his bosses, when he realized that this new stainless steel could be useful in spindles, pistons, plungers, and valves—in addition to cutlery. If anything, it was this tenacity—this quasi-insanity—that set him apart from earlier discoverers.

That same year, James Taylor returned from Australia, and moved in with Brearley and his wife for a year—no doubt soothing much of Brearley's fury.

Firth's, by then, recognized the industrial value of Brearley's steel for use in engine exhaust valves, and had begun marketing it as F.A.S.—Firth's Aeroplane Steel. In 1914, the company produced 50 tons of the steel; over the next two years, Firth's produced 1,000 tons more. Brearley bought 18 bars of it—125 pounds total, for 6 pounds, 15 shillings, 5 pence through Amalgam Co. (Two years later, this would have been illegal; with the Great War under way, the British government decreed that all chromium steels could be used for defense purposes only.) He made knives and gave them to his friends. He gave them to Stuart's friends. He instructed them to return the knives if, upon contact with any food, they stained or rusted. No knife was returned. Stuart knew he was looking at the future, and ordered, over a few weeks, seven more tons of the metal.

Success brought immediate animosity, because Brearley had one vision, and Firth's had another. Firth's omitted Brearley's name. Firth advertised itself as the discoverer, inventor, and originator of stainless steel, in ads, posters, and labels on bars of steel. One such ad, from 1915, said this:

FIRTH'S

"STAINLESS

STEEL"

for CUTLERY, etc.

NEITHER RUSTS, STAINS NOR TARNISHES.

———

Cutlery made from this Steel, being totally
unaffected by FOOD ACIDS, VINEGAR &c will
be found a boon in EVERY HOUSEHOLD and may
be had of all the LEADING MANUFACTURERS.

———

SEE THAT YOUR KNIVES OF THIS STEEL BEAR

THE MARK FIRTH STAINLESS

———

The daily toll at the knifeboard
or the cleaning machine is
now quite unnecessary.

ORIGINAL & SOLE MAKERS

THOS FIRTH & SONS, LTD., SHEFFIELD

Brearley complained and got a bitter letter in response. Yet he insisted. He told his boss, Ethelbert Wolstenholme, that he'd given Firth's a commercial opportunity, and that it had agreed to share any discoveries. But he'd also proven the company wrong and was to suffer for it. He was ignored, cast aside, told to deal with underlings. Annoyed and suspicious, he wrote a tactless letter to his boss. This led to a conference with Firth's three directors, who told him plainly that he had no rights in the matter. A few days later, on December 27, 1914, feeling wronged, more sad than angry— convinced that "workmen are often much wiser than their masters"—he resigned.

⌣⌐

Harry Brearley didn't know it then, but the stuff he cast from the electric furnace at Firth's on August 20, 1913, was nothing new. At least ten others had created it, or something like it, before; at least half a dozen had described it; and one guy even explained it, and explained it well. Others had patented it, and commercialized it. Before Brearley got around to it, at least two dozen scientists in England, France, Germany, Poland, Sweden, and the United States were studying alloys of steel by varying the amounts of chromium, nickel, and carbon in it. Faraday had tried as much nearly a century earlier. It's not like Brearley was exploring unknown territory. That he is credited with discovering stainless steel is due mostly to luck; that he is credited with fathering it is due mostly to his resolve.

Stainless steels, like whiskeys, come in many blends, and a Frenchman named Léon Guillet made five of them in 1904 but failed to notice their corrosion resistance. He made two more in 1906 and didn't notice them, either. In 1908 the German Philip Monnartz showed far more attentiveness. Describing his alloy in a paper three years later, he described a precipitous drop in corrosion with the addition of chromium, and called the phenomenon passivity. A year later, the German steelworks Krupp secretly filed patents for "fabrication of objects that require high corrosion resistance." Two metallurgists there had been examining nickel-chromium steel since 1908. One alloy they made is today the world's most popular. First called alloy V2A, it was marketed as Nirosta, and produced by the ton in 1914. But even that may not have been Krupp's first foray into stainless steel.

In 1908, Bertha Krupp, the wealthy daughter of the owner of the Krupp Works, commissioned a 154-foot-long steel-hulled schooner for her new husband, the Count Bohlen und Halbach. She spent $4.5 million (in today's dollars) lavishly outfitting the 191-ton *Germania* with white pine decks, Oregon pine masts, a 26-foot bowsprit, 15,000 square feet of the finest sails available, a stately dining room, and an elegant hull, painted white, made of chromium-nickel steel.

After the couple spent their honeymoon on board, an English crew raced the boat to victory at the 1908 Kaiser's Cup at Cowes, finishing fifteen minutes ahead of her nearest competitor, and setting a record around the Isle of Wight with an average speed of 13.1 knots. She made the kaiser proud. She won the race again, and others, before she was seized in Southampton on October 28, 1915, one of the first prisoners of the Great War. England had tons of a new stainless steel in her possession and didn't know it. Brearley, by then, was in possession of his own patent.

The early alloyers of steel had a difficult time commercializing their discoveries. Robert Hadfield, who made the first real, commercially valuable alloy, called Hadfield's manganese steel, had to wait a decade for it to take off. In his lab journal, on September 7, 1882, he described the peculiar nature of his alloy: it was soft but tough, and tempering it made it softer and tougher still; and even though it was 80 percent iron, it wasn't magnetic. It astonished him. He wrote, "Wonderfully tough, even with a 16lb hammer could hardly break it . . . Really grand. Hurrah!!!" It was bad for tools. It was bad for horseshoes. It was bad for fire pokers and car wheels. Hadfield grumbled in a letter to his American agent, "The material is being tried for a considerable variety of purposes but the people are so slow on this side & inventors here have so many prejudices." Finally, ten years after he made it, it was found ideal for railroad tracks. It lasted almost fifty times longer than carbon steel, and became the standard for heavy duty rails. With a new silicon steel Hadfield discovered in 1884, the situation was nearly the same. That time, he had to wait two decades before commercial viability became evident.

Commercial success demanded blending science and marketing; a steelmaker had to recognize not just the value of a new alloy, but its potential use. Benno Strauss, of the Krupp Works, later spoke about recognizing the potential of his stainless steel in plumbing, cutlery, medical equipment, and mirrors. He, like Brearley—who realized his stainless steel would be useful in spindles, pistons, plungers, and valves—was focused. He also knew that V2A wouldn't work. Nirosta would. Other steelmakers followed suit. Where the first alloys were named after the discoverer—Hadfield's manganese steel, R. Mushet's Special Steel, Firth's Aeroplane Steel (FAS)—later discoveries were marketed with an ear toward popular usage: Rezistal, Neva-Stain, Staybrite, Nonesuch, Enduro, Nirosta, Rusnorstain. In the naming of his alloy, Brearley owes a great debt to Stuart. He made FAS sound better than D.G.S. He made it sound like a miracle, and it worked.

A month after Brearley resigned, news of his stainless steel reached America. On January 31, 1915, the *New York Times* announced the discovery:

A NON-RUSTING STEEL;
SHEFFIELD INVENTION ESPECIALLY GOOD FOR TABLE CUTLERY.

According to Consul John M. Savage, who is stationed at Sheffield, England, a firm in that city has introduced a stainless steel, which is claimed to be non-rusting, unstainable, and untarnishable. This steel is said to be especially adaptable for table cutlery, as the original polish is maintained after use, even when brought in contact with the most acid foods, and it requires only ordinary washing to cleanse.

"It is claimed," writes Mr. Savage in the Commerce Reports, "that this steel retains a keen edge much like that of the best double-sheer steel, and, as the properties claimed are inherent in the steel and not due to any treatment, knives can readily be sharpened on a 'steel' or by using the ordinary cleaning machine or knifeboard. It is expected it will prove a great boon, especially to large users of cutlery, such as hotels, steamships, and restaurants.

"The price of this steel is about 26 cents a pound for ordinary sizes, which is about double the price of the usual steel for the same purpose. It also costs more to work up, so that the initial cost of articles made from this new discovery, it is estimated, will be about double the present cost; but it is considered that the saving of labor to the customer will more than cover the total cost of the cutlery in the first twelve months."

Here's another reason why Brearley is credited as the discoverer of stainless steel: Reporters at the *New York Times* weren't reading the metallurgical trade magazines. They didn't know about Monnartz and Strauss.

The first American ingot of Brearley's stainless steel was cast thirty-one days later and sent directly to a knife maker.

Not long after, a stranger named John Maddocks appeared on Brearley's doorstep. He was a well-dressed seventy-five-year-old from London, and he saw the future in stainless steel. He said he had experience getting patents and offered to get Brearley an American patent. First, though, Brearley had to determine the limits of chemical composition that gave his metal the properties it had. Which meant that he needed a lab. Firth's did not offer to help.

Brearley applied for a patent on March 29, 1915. He was denied, because stainless steel was being made in England by at least seven companies. He reapplied, specifying it as a "new and useful improvement in cutlery" for metal with 9 percent to 16 percent chromium, and less than 0.7 percent carbon.

It was a busy time for Brearley. In May he became the works manager at Brown Bayley's Steelworks; by the end of the year, the company, which had never successfully sold steel alloys, was making and selling crankshafts for airplane engines as fast as it could. Six months into the job, he was offered a seat on the board. He gave up the post after three years so that he could keep working on matters more compelling than business, like axles, springs, billets, and shafts.

Brearley was granted patent number 1,197,256 on September 5, 1916. By July 1917, he and Firth's had agreed to form the Firth-Brearley Stainless Steel Syndicate. Firth's paid Brearley £10,000 and half the shares for the interest in the patent, which was profitable, straight through the Depression. Brearley also got a bit of revenge, demanding that all knife blades made with the alloy bear his name, as Firth-Brearley Stainless.

Given the effort and assistance it took Brearley to commercialize his stainless steel, a quote from one of his business partners may be most apt. He said that Brearley knew everything about stainless steel except how to make it. What he meant is that Brearley knew everything about stainless steel except how to market it.

⌐‿‿‿⌐

At the end of that July, Elwood Haynes told Harry Brearley to hold his horses. Haynes, of Kokomo, Indiana, was a wealthy businessman, active in politics and public affairs, with a mustache like a windshield. He'd managed the Kokomo Gas Company, overseen the building of the country's first major long-distance natural gas pipeline, designed America's first true automobile and founded a company that sold them, and founded a company that sold an alloy he'd invented, called Stellite. Stellite had made him a millionaire. He'd also run for the US Senate (as a Prohibitionist), and a few years later was elected president of the YMCA. Haynes didn't want to sell stainless steel—he was too busy—but he wanted credit for the discovery.

Haynes had been experimenting with corrosion-resistant alloys for cutlery since Brearley was sixteen years old, reading algebra. He'd made chisels and augur bits and spark plugs out of chromium-steel alloys more than a year before Brearley had looked through his microscope and not seen what he'd expected. He'd also filed for a patent seventeen days before Brearley, which meant his claim had legal merit. The US Patent Office granted Haynes an interference order on July 31, 1917.

Haynes's patent, which was finally granted on April 1, 1919, was for steels with between 8 percent and 60 percent chromium—a huge range—but it didn't matter. Rather than litigate, the Firth-Brearley Stainless Steel Syndicate agreed to share profits, by forming the American Stainless Steel Company with Haynes and five steelmakers (Bethlehem, Carpenter, Crucible, Midvale, and Firth-Sterling, the Firth subsidiary in Pittsburgh) in 1918. For a decade and a half, business was stellar, even though stainless steel cost four times as much as carbon steel. From 1923 to 1933, American's average yearly dividend was 28 percent.

In 1920, when the Ludlum Steel Co. began selling stainless steel, American sued, and the court eventually held Ludlum accountable for profits from stainless cutlery. In 1933, when the Rustless Iron Corp. of America began selling stainless steel, American sued again—but this time the court detected no patent infringement. The court said that historically, "thousands of tests and experiments, and manifold achievements, had already made the metallurgy of iron and steel a broad and highly developed field. A claim to monopoly of any part of that field by one entering it at this late day can be sustained only by clear proof of discovery of something there not before found, of an invention of something not before there." The court, unlike the *Times*, knew about Guillet, Monnartz, Strauss, and others, thanks to Brearley, who testified. American appealed, and lost again. This time, the court said it more plainly. It said that the American Stainless Steel Company had a patent on a way of making a certain type of stainless steel cutlery, not all stainless steels used in all industries. That would be ridiculous.

Nobody knows if Brearley felt vindicated, but chances are that he enjoyed seeing Firth's greed stymied.

Brearley's greatest vindication, or validation—the climax of his career— came on a steely gray evening in 1920. It had been an unsettled spring: unusually mild in March and then unusually wet in April, and this night was the last of the dreary days, before a hailstorm the following day kicked off one of the sunniest Mays on record. That day, Thursday May 6, Brearley probably rode behind a horse on his way to London, to the fifty-first annual meeting of the Iron and Steel Institute.

The ceremonial dinner preceding the meeting was held at the Connaught Rooms, a palatial hall near Covent Garden, in the center of Westminster. The building, a stately five-floor affair twenty blocks north of Big Ben and the Houses of Parliament, is one of the most opulent in the city. From the main entrance on Great Queen Street, Brearley would have walked over harlequin tile, up twenty marble steps trimmed in brass, defended by columns, beneath an impressive chandelier. Ahead, passing through double doors beneath an arch, he would have entered a room that could have served as a train station, with forty-foot vaulted ceilings. It seated five hundred people. It was a far cry from any steelworks.

The dinner proceeded as most meetings do: with minutiae.

New members were announced. The previous year's annual report was read. The steel magnate Andrew Carnegie (who'd died the previous August) was mourned. To the son of a metallurgist who'd had a medal stolen during the German occupation of Belgium, a replica was given. The budget was reviewed. Library book donations were named. Finally, the retiring president of the institute introduced the new president, declaring that Dr. John Edward Stead had "rendered it possible to decipher those mysterious hieroglyphics which were impressed on steel by chemical reagents."

After much acclaim, Dr. Stead got to the big business of the night. He announced that the council had awarded the Bessemer Gold Medal—the highest award of the Iron and Steel Institute—to Brearley, "who, if not actually born in a steelworks, was cradled there."

The medal, for "outstanding services to the steel industry," particularly innovation in the manufacture or use of steel, was something Brearley had always coveted, on account of the recognition it implied. Isaac Bell, who built an iron works and Britain's first aluminum plant, and founded the Iron and Steel Institute, won the medal in 1874. Robert F. Mushet, who, as a result of over ten thousand experiments, perfected the Bessemer process, invented a heavy-duty rail alloy that was superseded only by Hadfield's, and created a self-hardening high-speed tool steel, won it in 1876. Frederick Abel, who studied the carbon contents of steel, the erosion of gun barrels, and all manner of explosives, won it twenty-three years later. Adolphe Greiner, the director of Belgium's most eminent steelworks, won it in 1913, and then had it stolen while he was taken prisoner by Germans during the Rape of Belgium.

Stead called Brearley devoted, the author of an excellent textbook, the brain behind a number of original researches. He said his work followed that of Faraday, and went through Brearley's discoveries. Then he said it was a great pleasure to welcome him to the fraternity of Bessemer medalists. All this in front of scores of his most eminent colleagues.

Brearley thanked the president, the council, and the members present, and said it was an additional honor to receive the medal from the hands of Dr. Stead. Then, in an agitated voice, he said:

> I have never lacked either good friends or kind counsellors; and I am sincerely grateful both to those who have encouraged me and to those who have opposed me. Indeed, I hardly know whether I owe more to those who approved or to those who honestly disapproved of the things I have tried to do. My brother was already engaged in the furnaces when I

began work; since 1882, we have not been separated. All the opportunities afforded to one have been shared by the other, and if it were possible to cut the Bessemer medal in two, I would give my brother half of it. No man ever had a better colleague, a severer critic, or a kindlier friend. Dependence on the willing help of others is part of all who successfully direct a laboratory for industrial research. Most problems relating to iron and steel cannot be definitely stated; and the individual, best qualified by experience to study the problem, and to solve it, might be some workman who was engaged on the job day after day. It is the investigator's greatest achievement to inspire interest in such men and make them confederates to his plan. The lust to work, the desire to find out and understand things is not confined to those who regularly wear clean collars; and, in thanking the Council of Members of the Iron and Steel Institute for this high honour, I am proud to confess my lifelong indebtedness to scores of friends, with hard hands and black faces, who toil at laborious tasks in mills and forges.

It was easily the most emotion-laden statement of the night, and also the funniest, for surely, if anybody knew how to cut a piece of metal in two, it was Brearley.

Afterward, Dr. Stead read a very long speech about blast furnaces, the puddling process, foundries, Bessemer and open-hearth processes, electric furnaces, the production of sound ingots, the application of science to the ferrous industries, and the advent and progress of metallography, among other things.

No photos were taken. Typical of engineers, the only visuals in the journal that records the event are photos of the microstructure of alloys, diagrams of furnaces, and graphs of temperature versus grain size.

Brearley retired in 1925. The year before, the world's most popular stainless alloy, type 304, was discovered by William Hatfield, Brearley's successor at Firth's. But he kept working, and was busier than ever. He helped J. H. G. Monypenny, the chief of the research lab at Brown Bayley's, write *Stainless Iron and Steel*, published in 1926—the first English book in the world on stainless steel.

Two years into retirement, Firth's dropped Brearley's name from Firth-Brearley without his permission. It made him furious and elicited the class warrior within. He yearned to level the field. When he wrote his memoirs, this bitterness was still potent, and gave him the opportunity to present to the world his only thoughts not directly related to steel. He thought directors should have to live within half a mile of their factories. He thought inheritance should be eliminated, along with dull work. He fumed that only the poor and inconspicuous get persecuted for bumbledom. Tired of being manipulated, he dreamed of meritocracy.

He figured there were four types of workers and four classes of wealth.

Workers
1. inspired and able
2. intelligent and diligent but not capable of improving
3. young promising lads
4. the half-asleep and half-blind.

Wealth
1. poor by luck
2. poor by will
3. rich by luck
4. rich by will.

He thought we should abolish the unfortunate first, honor the saints and philosophers of the second, choose wise men for the third (to which he belonged), and despise the profiteers of the last. Of how, exactly, this ought to be achieved—beyond Riga in 1905—he made no mention.

He proposed that steelmakers buy an abandoned steelworks in Sweden and turn it into a school for "budding manufacturers." The students would have shifts rather than schedules, and the place would run 24/7. They could sell the steel they produced. Grades would depend on its quality. Teachers—experts in aspects of metalworking—would be invited from all over the world.

In 1941 he founded the Freshgate Trust Foundation, which sponsored organizations making social, educational, medical, artistic, historical, and recreational endeavors. The way Brearley saw it, he wanted to "help lame dogs over stiles." But mostly, his naivete made him sound like a rambling idealist, a combination of Ayn Rand and Dr. Bronner. Like his steel, his ideas were born from his heart rather than his head. He had a vision of quality, or equality, but no idea how to get there.

These were just ideas; the most action Brearley ever took related to his son's education, and even then he capitulated. He'd refused to send his son to school; he believed that education ought to be chosen, rather than imposed, and considered the school system "an unloading of useless information into developing minds interested in other things," turning children into parrots. He told a few members of the school board that he wasn't sending his son to school because he wanted him to grow into an educated man. He figured that mass education produced a "kind of industrial unit, obedient, submissive, lacking in healthy curiosity and initiative." The school board members said his son would have to be tested periodically. He questioned the worthiness of the test standards, and refused to compromise. The episode seemed destined to involve the police—and then Brearley was sent to Riga, and relented. That was two decades earlier.

The only action at the end of Brearley's life consisted of rolling bowling balls, spinning tops, collecting stones, and playing marbles. Ashamed at the quantity of marbles he took from kids, he gave them back for free, only to win them again. He never handicapped himself. He tried gardening but couldn't remember the names of all the growing things, even with labels. It just didn't stick. He relaxed. He read Whitman, Shaw, and finally Shakespeare. He listened to Handel. He wrote his autobiography, as well as an unpublished children's book, called *The Story of Ironie*.

Actually, there was one more symbolic action. Before going on his first vacation (beyond Riga, he'd been to New York to serve as a witness, and Berlin for a conference on applied chemistry), at age fifty-eight, he piled up all of his index cards and a few hundred books and burned them in a great bonfire. Perhaps it seemed a fitting ritual to honor his mentor James Taylor, who'd recently died. Perhaps, tired of all the litigation pertaining to stainless steel, he lashed out in that strange way. He'd never been much of a servant to pencil and paper. More likely, he felt that such knowledge, in the changed world of steelmaking, was no longer relevant.

His class warfare turned to paranoia. On the last day of 1931, he left a sealed envelope, with instructions not to open it until 1960, with the

Cutler's Company, in Sheffield. The packet contained his sworn declaration on the history of stainless steel, implying some ugly secret best disclosed when no harm could come to him or others. The revelation was a dud; it revealed more about the man than about the metal.

The next year, Brearley was thinking about posterity. Sorby's name lived on, in the form of small (£12) annual memorial prizes. Brearley decided to honor James Taylor the same way, by endowing a biennial prize for the same amount in his name, through the Sheffield Metallurgical Association, for the most original or useful metallurgical paper.

His son, Leo, and his brother Arthur both died in 1946, two years before him. Stainless maybe, but not unkillable. Harry Brearley died more than three years after V-E Day—which means that he got to see the metal he fathered contribute to the fighting of another brutal world war. Of the destructive power that his hardened steels contributed to, he made no comment. On his service to the war machine as a veritable military contractor—during his career, he improved on armor-piercing shells, battleships, armor plates, gun barrels, and crankshafts for air force planes—he said nothing. Then again, in his autobiography, he doesn't mention his wife, his son, or anything funny, much beyond steel. The rebel himself was almost as cold as steel.

The Iron and Steel Institute recorded his death as it did that of any other member of the industry. From its August 1948 volume:

"The Council regret to record the deaths of:

"Mr. Harry Brearley, on the 14th July, 1948, at Torquay, aged 77."

Two pages of announcements of meetings, scientific advisory councils, refresher courses, and student exchange programs precede an obituary—a full column and a quarter—written by J. H. G. Monypenny, coauthor of the stainless steel book.

"By the death of Mr. Harry Brearley," he wrote, "the iron and steel industry has lost one of its outstanding personalities."

Monypenny remembered Brearley as an apt pupil, a keen observer, and said he had "a lucid style of writing." He went on: "throughout his life he was an individualist and professed to have little use for Technical or Research Committees . . . He had little respect for tradition and often attacked it in his writings. In his last book . . . he attacked, in characteristically iconoclastic vein, a number of traditional views on certain aspects of steel metallurgy which he particularly disliked . . ." And then, craftily: "As a man he was kind and helpful to those who, he thought, needed his help." Unwritten: most of the time, he was a stubborn, rebellious, opinionated curmudgeon.

The index in that volume of the journal, though, would have made Brearley proud. It lists thirty-seven references to stainless steel—one of the longest in the entire index—among them: arc welding, bright annealing, centrifugal casting, characteristics, classification, cold rolling, condition-ing, cutting, descaling, drawing and pressing, ductility, forging, passivation, pouring, properties, self-soldering, and welding. (And one that Brearley would appreciate: "applications in America.") His work lived on, and was thriving.

One paragraph from Brearley's memoir, though, describes his outlook best:

> The range of the mind's eye is restricted by the skill of the hand. The castles in the air must conform to the possibilities of material things—border-line possibilities perhaps; or, if something beyond the known border is required, the plan must wait until other dreams come true. If the plan works, the myriad possibilities of the vision, which became an idea and then a selected instruction, become ultimately an object. A vision crumpled into a thing. The Word made Flesh. Something for all mankind to take hold of, to use, to look at, to wonder over, and to utilise as a stepping-stone to other visions.

4

COATING THE CAN

In the quest to design the perfect beverage container, what you don't want the container to do matters as much as what you do. You don't want the container to impart any flavors to the beverage within. You want it to be cheap. You don't want it to be heavy, but you do want it strong and durable and stackable. Also, however minuscule the chance, you don't want the container to explode. In this regard, aluminum cans outperform glass bottles.

Since 1911—when Sam Payne, in Rome, Georgia, lost sight in one eye when a bottle of Coca-Cola exploded and the court found fault with the bottle maker—victims of exploding glass bottles have included everyone from infants to housewives. Waitresses and grocery store customers have suffered greatly, mainly with injuries to their appendages. At least one victim of an exploding bottle has endured a severed Achilles tendon. Most common, or at least most often litigated, and most frightful, are eye injuries. An exploding bottle will take an eye out. It doesn't take a JD to recognize the liability to a company in the beverage-container industry.

The beverages inside exploding bottles run the gamut. You name it, it's exploded: soda, beer, champagne, Perrier, grenadine, milk. Bottles have exploded in bars and restaurants, hardware stores, drug stores, and liquor stores. They've exploded at Stop & Shop. They've exploded at Safeway.

Bottles have exploded in the parking lot, in the car, in the kitchen, in the garage, and in a dorm room. Bottles have exploded during the walk home, and at a picnic. They've exploded while putting bottles in the fridge, while putting bottles in the pantry, while transferring bottles to a cooler, while transferring bottles from a cooler, and while not transferring them at all. They've exploded during delivery and months later. One case sounds like something out of a James Bond movie: room service delivers a soda to a hotel guest. It explodes in his hands. Another sounds like something out of the *Onion*: an employee at Pepsi loses an eye when a Pepsi bottle explodes. Injured parties have described the sound of an exploding bottle as like a shotgun, or a lightbulb dropping, or a firecracker.

At least 130 such cases have gone to trial. One plaintiff asserted that there were at minimum 10,000 exploding bottle episodes annually in the carbonated beverage industry. As the court stated in 1961, the danger that a glass bottle filled with Coca-Cola under pressure may explode is obvious. Bottle makers, desperate to demonstrate that exploding bottles have been dropped, kicked, abused, mishandled, tampered with, run over, treated negligently, or left rattling around in the back of a pickup truck for a summer day, dream of the calm litigious waters at the offices of their can-manufacturing rivals.

Yet those waters aren't entirely placid. In 2008 Lynda Ryan, of Blythe, California, lost vision in her left eye when a can of Diet Pepsi exploded as she was opening it. On the day it happened—July 9—the temperature reached 109 degrees, a few degrees shy of the record. It didn't get much below 90 at night. The Diet Pepsi was in a cooler in the back of Ryan's van, and had been there for several days. The temperature inside her vehicle was probably near that of the sun. The company could certainly argue that no respectable, let alone perfect, aluminum can could survive those conditions. She settled confidentially out of court.

Other exploding can incidents never made it to court but did make their way into the record books of the Consumer Product Safety Commission. A sample of some entries:

Urbana, IL: soda can explodes in fridge, cuts thumb of consumer.
Undisclosed location: soda can explodes in hand of consumer. No
treatment necessary.

Waldwick, NJ: beer can explodes in fridge, lacerates hand of consumer.

Zebulon, NC: two soda cans explode on top of fridge. No injuries reported.

Undisclosed location: beer can explodes while loading a cooler, lacerates nose of consumer.

Red Bank, NJ: three soda cans explode inside fridge. No injuries reported.

Undisclosed location: soda can explodes in minivan, at the hand of twelve-year-old. Child is unhurt. Dad drives into guardrail.

As a result, new cans made by the world's largest can manufacturer contain an "anti-missiling" feature. It keeps the panel from blowing out and getting you in the eye.

Also relevant in the quest to design the perfect beverage container is the likelihood that a foreign object may find its way into the container. In this regard, aluminum cans underperform glass bottles. Between the time that a two-piece can is made and the time it is filled and the end is fastened on lies a vast opportunity for objectionable things to find their way in. These things could be as simple as beans, peanuts, dirt, or pine needles, or they could be lightbulbs, bells, bobby pins, paper clips, tacks, safety pins, matches, film, AA batteries. These have all been found in drinks, and pose similar problems of liability.

Like exploding bottles, these episodes also find their way to court. Sometimes the cases involve cigar butts, cigarettes, a condom, a tampon, or a bandage. One case involved a cigar *and* an insect, in the same drink. Many cases involve slivers and flakes of metal, or bits of rust. Plaintiffs have won cases for drinks containing ants, bees, centipedes, cockroaches, flies, grubs, maggots, moths, worms, roaches, a yellow jacket, a wasp nest, a black widow spider, and a snake.

Plaintiffs have found in their drinks: a roach egg, "a partially decomposed worm or cocoon," insect larvae, a fly covered in fungus, a putrefied mouse, a mouse skeleton, "a rat with the hair sucked off," a piece of flesh, something identified as part of a mouse or bird, rotten old meat, and "blood vessels of unknown origin." In three dozen cases involving dead mice, the court has found for the plaintiff. Plaintiffs have vomited blood, excreted blood, had ulcers and dysentery. Many have been bedridden for weeks or felt "deathly sick." Many have been hospitalized. One had stomach surgery; another lost thirty pounds in five weeks. Most never drink soda again, out of the perfect container or otherwise.

It's hard to say how many of these cases are legit, because there's also the story of an Illinois man named Ronald Ball. Mr. Ball claimed that on November 10, 2008, he bought a can of Mountain Dew from a vending machine on the outskirts of St. Louis, opened it, took a sip, spit it out, vomited, then poured the rest of the can's contents into a Styrofoam cup, whereupon he discovered a dead mouse. He called the number on the can, complained, and, as requested, sent the mouse and the remaining Mountain Dew to PepsiCo, its manufacturer.

From the serial number on the can, Pepsi determined it had been filled seventy-four days before Mr. Ball opened it. To assess the mouse, Pepsi sent it, in the Mountain Dew, in a glass jar, to a veterinary pathologist in Salt Lake City named Lawrence McGill. McGill, who had conducted thousands of necropsies, and claimed familiarity "with the effects of an acidic fluid" on mice and other animals, opened up the jar, and put the mouse in formaldehyde, to arrest its decomposition. The next day, he cut open the mouse, and went hunting for clues. In the mouse's leg and head, he found bones; in the mouse's unruptured abdominal cavity, he found a liver, intestine, and stomach; and in the mouse's lung, he found cartilage cells—indicating that the mouse had been in the Mountain Dew no more than a week. Because he was unable to open the mouse's eyelids, he deduced that the mouse was young, at most four weeks old. The Mountain Dew, he determined, had a pH of 3.4. The conclusion McGill reached was simple, based on unanimous, irrefutable evidence: the mouse did not exist at the time the can was filled and sealed, and had not been stewing in Mountain Dew for more than a week, let alone seventy-four days. The pathologist said as much in a signed, notarized affidavit dated April 8, 2010.

Three weeks later, by which time Mr. Ball complained that the dissected mouse (which had been returned to him) had been rendered unfit for further testing, he sued for $325,000. Pepsi's lawyers got straight to the point. Citing McGill's findings, they said that Mountain Dew was so corrosive that, had a mouse been subjected to the concoction for seventy-four days, as Mr. Ball alleged, it would have dissolved into a "jellylike substance," leaving no identifiable mouse parts behind. The judge vacated that case.

Mr. Ball sued again, seeking $50,000 in relief. The judge dismissed that case too. Before it was all over, the news-of-the-weird story was on every channel in America. Eric Randall, in the *Atlantic*'s blog, summed it up best. He wrote, "This seems like a winning-the-battle-while-surrendering-the-war kind of strategy that hinges on the argument that Pepsi's product is essentially a can of bright green/yellow battery acid."

Therein lies the crux of designing the perfect beverage container: either the container does not want to cooperate, or the beverage makes containment difficult. Or both.

In the case of the aluminum can, it's both and then some. For the country's largest can maker, the aluminum that is mined in Dwellingup, Australia, and smelted in Evansville, Illinois, and manufactured in Golden, Colorado, into twelve-ounce cans behaves throughout the can-making process begrudgingly. It tries to jam the machines that bring it into existence, and strays toward a number of disastrous scenarios: it may stretch, crumple, fracture, collapse, pleat, buckle, or blister. If it emerges as a healthy can, it still aspires to misbehave, by refusing to protect whatever beer, soda, energy drink, or other "product" ends up inside it. It wants to interact with the product, and change its taste. Worse, the can still finds more ways to throw a tantrum. Beyond exploding, it may leak, or somehow corrode: from the inside out, or the top down, or the bottom up. Rust is a can's number one enemy. Manufacturing strong, healthy aluminum cans, in fact, is so challenging, and requires such a vast amount of study, design, and precise machining, that many consider cans the most engineered products in the world. This notion—that the ubiquitous aluminum can, which seems anything but amazing, is in fact incredible—was the first thing I learned at Can School.

The second thing I learned at Can School is that of all the operations used to stave off aluminum's suicidal tendencies, wheedle the metal into submission, and avert what more than one can-making employee called time-bomb behavior, the corrosion-related procedures are so sensitive and shrouded in secrecy that asking lots of questions about them is a good way to get kicked out of Can School.

Can School was put on by America's largest can maker. Over three days, in the spring of 2011, engineers, chemists, and managers from the company

discussed "improved pour rates" and "recloseability" and the overall "experience from a can." Except they didn't call the common 12-ounce can a can.
They called it a 202 (because the diameter across the top is 2²⁄₁₆"). Most
wore cell-phone holsters, and many sported mustaches. One described
cans as "like sunshine"; another discussed the "opening performance," and
didn't mean the opera. Nearly sixty attendees—from Heineken (Mexico),
MillerCoors, Nestlé, and Pabst—listened attentively at four long tables in
a semicircular conference room just north of Denver.

On that first day—before I nearly got kicked out—to my right sat
three women from Pepsi, with New York accents. To my left sat a man
from Anheuser-Busch, who told me that he hedges a year out on the London Metal Exchange on $1.5 billion of aluminum annually. Left of him sat
a guy from the Dairy Farmers of America. I heard them conferring about
Americans' milk-drinking habits. In front of me sat two guys from Coca-
Cola. Behind me sat three more guys from Coca-Cola. One attendee wore
a shirt with a patch that said "Can Solo." Another gave me a business card
with a picture of six canned beers and his title: Can Whisperer.

Displayed prominently on a screen was the motto "eat. drink. imagine."
On the left wall was a table covered in various food cans, from Crisco to
Chef Boyardee. On the right, various beverage cans: Molson, Labatt, Fos-
ter's, Pabst. At my seat: a black folder, and two posters, showing the steps
of the can-manufacturing process. There was nothing rusty about it. Beside
the screen stood a podium, and to its left, an American flag. To its right
stood the baby blue flag of the Ball Corporation.

If you drink beer, or soda, or juice, or water, or sports drinks, or coffee, or
milk, or, really, anything, or if you have ever preserved fruits or vegetables in
glass jars, you may recognize the name. Grab a can of beer and start hunting. The Ball logo—cursive, underlined, and slanted upward—is minuscule,
but it's probably there. On a Pabst can, it's about an inch below the rim,
above the bar code, above the sell-by date, just above the government warning, just right of the word *problems* in "may cause health problems." But it's
not on every can; it's up to the product manufacturer to decide if it wants
the Ball logo there at all. It's on Miller Lite, Stroh's, Heineken, Schlitz,
Miller High Life, Tecate, Colt 45, Blue Moon, Honey Brown, Stella Artois,
Dr Pepper, Mountain Dew, Pepsi, Coke, Schweppes, Izze, and Starbucks,
but it's not on Monster Energy, Budweiser, or Bud Light.

The Ball Corporation has been running Can School annually for twenty-five years; not quite a thousand people in the beverage industry have graduated. To say the company is qualified to teach the course is a gross understatement. The people of the world go through 180 billion aluminum beverage cans a year. That's four six-packs for every person on the planet. The United States and Canada gobble up more than half of them—100 billion a year—and Ball makes a third of these. (Two other companies make the majority of the rest.) Ball, which has a long history with containers, operates thirteen beverage-can plants in Europe, five in China, and five in Brazil. It also operates fourteen steel food can factories in the United States and one in Canada. In pursuit of making cans, Ball employs fourteen thousand people. The total area of just the company's American beverage can factories approaches six million square feet. Ball is the third largest manufacturer in China, the second largest manufacturer in Europe, and the undisputed major player in the US beverage can industry. The company makes about a quarter of the world's beverage cans.

Since Ball seriously entered the can market, in the 1980s, its stock has outperformed DuPont and American Express, and trailed just behind ExxonMobil. Since 1994, when Ball first made $1 billion worth of cans, its compound annual growth rate has exceeded 12 percent. Since 2002, when Ball bought Schmalbach-Lubeca AG for $1.18 billion, it's held the title of largest can manufacturer in the world. In 2009 the Fortune 500 company acquired four more US plants, in Georgia, Ohio, Florida, and Wisconsin, for which it paid more than half a billion dollars, and as a result of which its market share increased to 40 percent. That same year, Ball sold $4.6 billion worth of aluminum beverage cans, earning $300 million in profit. Ball makes satellites, too—but they're not the big moneymaker. The big moneymaker is beverage cans, at a dime apiece. Sell forty billion of them—earning two-thirds of a penny on each—and you'll make out better than the Dow Jones Industrial Average and the S&P 500. But that doesn't mean it's easy.

⌒

On account of corrosion, it took engineers 125 years of tinkering with steel can designs before they figured out how to encase beer within, another quarter century for them to wheedle aluminum into service, and most of another decade for them to swap the beer with Coke. Consider a can of Coke. It's a corrosion nightmare. Phosphoric acid gives it a pH of 2.75, salts and dyes render it still more aggressive, and the concoction exists under ninety pounds per square inch of pressure, trying to force its way out of a layer of aluminum a few thousandths of an inch thick. It sits there for weeks, months, years, often in a humid fridge, or dank pantry, or hot trunk, or stagnant warehouse. That the can doesn't corrode is a technological marvel. That we are capable of reproducing that result hundreds of billions of times over—with a failure rate of 0.002 percent—is an unheralded corrosion miracle. And Coke is just the beginning. In the near half century since it was first canned, we've packed more corrosive beverages, such as San Pellegrino, V8, and Mountain Dew, into cans that have concurrently gotten thinner and more delicate.

All that protects the meager aluminum is an invisible plastic shield. Industry insiders call it an internal coating, or IC, and it's the product of a phenomenal amount of work. This plastic must be tough but flexible. It also must be rheologically suitable in terms of viscosity, stability, and stickiness. Without this epoxy lining, only microns thick, a can of Coke would corrode in three days. Our stomachs are stronger than the aluminum. But it's becoming increasingly evident that other parts of our bodies may not be stronger than what's in the epoxy. That's why the beverage container industry would rather not talk about rust, and why I nearly got kicked out of Can School.

⁓

Before coating the insides of their cans, Ball needs to know how corrosive the product—the beverage within—will be. The epoxy coating, after all, isn't free—it costs about a half penny per can—and Ball doesn't want to waste it. Also, some beverages are so corrosive that no amount of coating will protect their cans. Ball is not in the business of sending cans out into the world to be slaughtered by overaggressive liquids. The coating must perform. Otherwise, cans explode, and legal costs climb.

Ed Laperle, a tall, thin corrosion engineer at Ball, told me that until twenty-five years ago, Ball solved this problem by studying test packs. Five months before Can School, Laperle gave me a tour of the company's corrosion lab, which it calls its Packaging Services lab. Laperle, whose white beard makes him look astonishingly like Bob Vila, told me how the old system worked. Until the 1990s, he would tell customers, "Okay, send me some, I'll put it in a can, on a shelf, for six months and see what happens." Cases just sat there. Customers had to wait. Eventually Laperle figured out whether the beverage demanded a can with a windbreaker or a down jacket, and called back and said, sure, we can put your product in our cans. It didn't take long for customers to tire of the wait. Scott Brendecke, a corrosion engineer who works for Laperle, described the pervading logic that evolved. Customers cited historical performance, hoping to weasel out of test packs. "People would say, 'Oh, this root beer is just like that cola,'" Brendecke said. Like any good engineer, he disliked the logic. "The category of 'similar' just broadens over time," he said. "There's no learning. Eventually failure." By failure, Brendecke meant leaks and explosions.

Then Ball figured out how to do the corrosion test in about four hours. They call it a pitting scan. Using a potentiometer the size of a desktop computer, and a simple wire-frame setup that looks like it cost $100 at Radio Shack and would be suitable for a high-school chemistry lab, the engineers in Laperle's lab apply a tiny DC current to a sliver of encased aluminum sitting in the liquid in question, which sits in a glass jar. Over four hours, the potentiometer spits out a graph of current versus time. It looks like a pyramid. The peak represents the pitting potential, or PP, of the liquid. The pitting potential reveals how much current is required to remove the outer layer of aluminum oxide from the aluminum sliver. Once that outer layer is removed, corrosion can proceed unabated.*

By plugging the pitting potential and some other measurements (salt, copper, chloride, dye, dissolved oxygen, pH) into a carefully guarded equation, Laperle's team can determine the corrosivity of the product. The corrosivity, in turn, determines the thickness of the coating on the can it'll be in. Beer, for example, isn't very corrosive, so coatings on beer cans are extremely thin, and weigh in the neighborhood of 90 milligrams. Beer just happens to be amenable to coexistence with aluminum, because of its mild pH and some other convenient traits, which I'll get to later. Coke, which

---

*Pitting potential studies were pioneered by a giant in the field, Herbert Uhlig, who started the corrosion lab at the Massachusetts Institute of Technology that now bears his name, and who also wore a swarthy little mustache.

is more corrosive, demands a heavier coating. Especially acidic beverages, like lemon-lime drinks, and salty, or "isotonic," drinks, such as V8, demand greater corrosion protection, and hence thicker coatings of up to 225 milligrams. The coating thicknesses are referred to as *A*, *B*, and *C*. Neither Laperle nor anybody else in the can-making industry, though, would tell me exactly what they are. The most anyone revealed is that the average is 120 milligrams per can.*

Laperle has come across many products that are too corrosive to put in a can with even the thickest coating. If a beverage's corrosivity is over the line—if, say, it's a particularly potent "bright green/yellow battery acid"— Laperle will tell the product manufacturer that the beverage needs to be modified before it can be put into a can. When Laperle does this, over the phone, he is direct. "I tell 'em straight up, 'Your product failed. You can reformulate. Here are some suggestions.'" He'll advise raising the pH or cutting down the dyes. He has figured out that if he gives the manufacturer a goal, the conversation will go more smoothly. "If you just tell 'em 'It failed,' they hang up unhappy," he said. Specific cases he refused to cite. A few times, with products that were an order of magnitude too corrosive to put in cans, no matter what Laperle tried, he called the manufacturer and said, "We're kinda done. There's nothing else I can do for you." Yes, the can wags the product.

This much is clear: the higher the pitting potential of a beverage (anywhere from 100 to 500 millivolts), the fewer cans will fail. This much is also clear: sodium benzoate is bad. Copper is bad. Sugar is good. It absorbs carbon dioxide, decreasing the pressure within a can, and it also inhibits other corrosion reactions, because sugar tends to deposit onto pores in the coating. Thus, Diet Coke underperforms regular Coke on at least two counts. Citric acid and phosphoric acid are equally bad. Red #40 is pretty bad, and high chlorides are really bad, and together, they're really, really bad. At Can School, degrees of badness were represented on a graph of pitting potential versus chlorides, with a line asymptotically falling from

---

*If you happen to have a fume hood and half a liter of sodium hydroxide, you can expose the epoxy coating by dissolving the aluminum. Alternatively, if you fill a can with a solution of copper chloride, having scratched the coating first, after a half hour you can tear it open and peel the coating out.

the comfort zone, through the worry zone, and into the extreme worry zone. The extreme worry zone is where failures happen, where cans corrode from the inside out. When that happens, you get explosions and then lawsuits.

Brendecke later explained internal corrosion with a good metaphor. He held up two hammers, one in each hand. In his right hand, a metal hammer; in his left, a toy hammer, made of inflatable purple plastic. "You can bang on your beverage can all day," he said, swinging his left hand, "and nothing will happen." Or, he said, you can add corrosive elements, until the weight of the head of the plastic hammer resembles that of the metal one.

But the pitting potential is measured in a jar in a lab, rather than in an actual sealed can out in the world, so it's not absolute. That's why the corrosion engineers do a few more tests, to examine the package-product interaction, or PPI, of the coating and the beverage. As in the old days, they do test packs, storing eight cases for at least three months, so that they can determine if any of the products are absorbing any metal. Metal pickup must be less than two parts per million. They use a spectroscope to check. Laperle and his staff also test their coatings, using electro-impedance spectroscopy, or EIS. To do this, they modify the pitting scan setup a bit. Instead of using a sliver of uncoated aluminum, they use a two-inch square of coated aluminum. Instead of using a beverage, they use a salty and acidic corrosive liquid called liquor 85. And instead of applying DC current, they apply AC current across a range of about forty frequencies, from about 100 kilohertz to 10 millihertz. The resulting relationship between voltage and current, measured over a couple of days, reveals the values of resistors and capacitors in a model of the various electrochemical reactions taking place. Modeling this impedance, which involves deconvoluted voltages and non-faradaic components in the inner and outer Helmholtz planes, takes most of a PhD in chemistry. Calculating it takes a good potentiostat. The result yields the strength of the coating.

EIS has been around for a century, but didn't take off until the 1970s, when potentiostats became reliable. Since then, the technique has been used to study everything from semiconductors to protein reactions. Ingeniously, the technique was perfected over years and then taught, during a weeklong course in Thorton Hall, by a pair of professors at the University of Virginia. Ray Taylor designed, and still privately teaches, the annual course. John Scully was one of Taylor's lecturers. Taylor now runs the National Corrosion Center, at Texas A&M University. Scully, still at UVA and by his own admission on a corrosion crusade, has since become the editor of the journal *Corrosion*.

Laperle didn't take Taylor's course, but his colleague Jack Powers did. He was Ball's manager of chemistry. This was in 1987. It was the first EIS course that Taylor taught, and among the thirty-two attendees were as many employees of battery companies as employees of can-making companies. Taylor, not recognizing the animosity between the makers of steel cans and aluminum cans—they can hardly stand to be in the same room—had invited both. Powers, though, must have liked what he saw, because he contracted Taylor to research corrosion issues in cans. "People don't realize that the survivability of a can is almost entirely a product of its lining," Taylor told me. He was contracted to develop a test that could assess that survivability in a week. "It's really a brutal industry," he recalled. With test packs, "You've gotta wait twelve months to find the answer. By that time your competitor has surpassed you, and you're stuck there waiting." He continued: "The margins are just ridiculous. They've gotta make a billion things, and they've gotta be perfect. It's just mind boggling." Soon enough, Taylor reported to Ball that, in a blind test, his quick newfangled results rivaled Ball's slow old-fashioned results. A quarter century later, they're still the standard.

These days, employees of Ball's Packaging Services lab test fifty new products a month for corrosiveness. This is four times the work they did a decade ago, mostly on account of what Laperle called mom-and-pop shops, selling "Bob's Energy Drink," which was most likely created by a flavor house unfamiliar with corrosion. According to Laperle, chances are about one in seven that Bob's Energy Drink, created this way, will fail. Major suppliers such as Coke and Pepsi, on the other hand, claim a perfect, untarnished record—"bright green/yellow battery acid" notwithstanding.

Ball doesn't manufacture cans by the billions until after the beverage and the coating and the interaction between the two have been examined. After the company's engineers examine package-product interaction chemically, they double-check with their tongues. After all, you don't want your perfectly designed container to impart any flavors on the beverage within. For these studies, there's the flavor room. During Can School, Laperle gave a tour.

The flavor room sits across the hall from the corrosion lab. The cabinets on one wall made it look, ostensibly, like a kitchen, but on the way in, I passed a door with a chemical hazard sign: 4s all around. Full of extremely hazardous concoctions, it was no broom closet. There was also a minifridge, with a bar tap, and forty tiny brown vials on a table. The vials contained six scents that required the full range of nasal observational powers. Each was denoted by a little colored dot on the lid. Laperle, who has a quiet, scratchy voice, instructed me to name the smells. I slid a handful of vials in front of me, and began sniffing. The first was a mystery. I recognized it, but couldn't nail it. As I sniffed, Laperle said there was no such thing as a container that doesn't change the flavor of the product. Cans, plastic, even glass, he said, had an effect.

As Laperle explained the need for flavor testers, judging various PPIs, I began sniffing a second vial. It also seemed familiar, but locked away in a remote part of my scent-memory. The third was even more familiar: it turned out to be pine, but someone else said it before I figured it out.

Neither boasting nor bored, Laperle stood in the corner. He said that to be eligible for hiring as a flavor tester, I'd have to get seven out of ten such samples right. After that, it would take eighteen months to train me. It turned out that the vials contained almond, banana, and an antiseptic that smelled like Band-Aids. I didn't get a single one.

Laperle said that flavor testers at Ball learn to detects parts per million, then parts per billion, and eventually, parts per trillion. He figured that if his flavor testers couldn't taste something, nobody could. The one exception may be cats. On account of felines' extreme organoleptic capacities, wet cat food is packaged in cans with "particularly low levels of taint." (This, according to the book *Metal Packaging: An Introduction*, by an English beverage-can consultant named Bev.) Laperle, who has a dry sense of humor, later referred to this particular talent of cats as a type of idiot savantness, but I bet that if cats drank beer and their meows could be interpreted, Laperle would hire them.

Refining nasal senses is hard, Laperle explained, because we associate smells with experiences. For a while, he said, he'd do it at home over dinner, to his wife's annoyance. "Oh, I smell ketones, aldehydes, and fatty acids," he said, widening his eyes. Laperle made no wild or quick gestures; he moves deliberately. He said he did it at restaurants, scribbling on napkins. He said eventually you teach yourself how to turn it on and off, to enjoy eating without getting all technical about smells.

Laperle started working at Ball in 1980, after earning a master's degree in microbiology from the University of Massachusetts. He started the corrosion lab that he now directs. It's not like corrosion was calling him, though. Before graduate school, he'd worked as a cook, as a carpenter, and in a few factories, doing what he called "numerous and sundry things." For his bachelor's degree, he studied food science, because he thought he wasn't smart enough to be a chemical engineer. As he stood in the corner, he seemed comfortable, amiable, like a proud father. He suggested that Ball start the lab after growing frustrated by slow-working consultants who charged millions of dollars to solve the mysteries of Ball's field failures—leaks or explosions. The lab was born in 1983. Since then, Laperle's learned a lot about subtle flavors, especially in beer. Laperle knows what he's talking about; his tongue can detect one part per million of oxygen. He's a judge at the Great American Beer Festival. I once asked him if he had any favorite beers. His favorite, he said, is the one he was currently drinking.

Describing unappealing flavors, he said that the bad notes are always hidden on the gas chromatograph behind other bumps. He's better than the machine. He said that beer was very susceptible to "flavor scalping" if it comes into contact with too much coating. The cost alone persuades Ball; the smell just rubs it in. Beer, Laperle explained, is actually so mild that the can does not require a coating. He called beer a "nice oxygen scavenger," describing how proteins in beer consume dissolved oxygen, keep it from accessing and corroding the aluminum. It's the same for orange juice, in which vitamin C consumes oxygen—which is why canners were able to package it so long ago. It turns out that cans were made for beer, and beer was made for cans. In fact, the only reason beer cans have a coating at all is so that the carbon dioxide doesn't escape at once. The coating smooths out the surface of the metal, so that the gas has no microbumps from which to propagate, as it does on beer steins designed for that purpose. Nobody wants a can of flat beer. The coating keeps it tasty.

And if the taste of "bright green/yellow battery acid" is particularly appealing to you, the coating tested in Laperle's flavor room also deserves some credit.

⌒

The name *Ball* probably makes you think of glass jars. Technically, they're Mason jars, stamped with the name Ball. Your mother probably had some in the pantry.

Ball jars go back to 1882, when the five Ball brothers—Frank, Edmund, George, Lucius, and William—started making glass jars in Buffalo, New York. For marketing purposes, they began growing their mustaches shortly thereafter. Both took off.

Within five years, they were making more than two million glass jars a year. They relocated to Muncie, Indiana, and with natural gas rather than coal, figured out how to quintuple production. By 1893, they had a thousand employees. In 1895 they made 22 million fruit jars. The next year, they made 31 million. The year after that, 37 million. In 1898 the brothers patented a semiautomatic glass-blowing machine, which tripled productivity. Two years later, they invented the first automatic, a.k.a. electric, glassmaking machine, which meant they could crank out seven times as many jars as they had a dozen years before. They were limited only by how fast they could bring glass to the machines, and in 1905, they automated that, too, nearly doubling production. By 1910, they were producing one jar per capita. That was 90 million jars. The jars were stored outdoors, on a field, in long rows, leaning sideways. The field was so vast that migrating ducks tried to land there, mistaking the glimmering glass field for a lake.

As business grew, so did their mustaches. Frank's mustache was thick, and covered his whole mouth, Teddy Roosevelt style. George's was thin, and tapered to fine points, like a French connoisseur. Lucius had handlebars, extending to his ears. Edmund had the humblest mustache, à la Pancho Villa. William had my favorite: a horseshoe connecting to the lower half of a beard, and long, straight sideburns. As they became successful, each, in his own way, revealed panache, bravado, flamboyance, and fortitude.

Business was phenomenal. Food canning took off during the Panic of 1893, and bloomed between the Great Depression and the Second World War. It was good for the country. It seemed the brothers were making the perfect container. The jars were tapered, rounded, square, rounded square, tall, and squat. The glass was amber, aqua, clear, blue, yellow, green. The brothers built a rubber plant. They bought a zinc mill, moved it to Muncie, and expanded it until it was the largest rolling mill in the world. To ship their jars, they bought a paper mill, and then two more. They already owned a railroad company. The brothers all built mansions—William's a Georgian, Edmund's a Tudor, Frank's a Victorian—on the same boulevard in Muncie. They founded Ball State University.

By 1936, the Ball brothers had 2,500 employees and were making 144 million jars a year—more than half the country's fruit jars. Ball displayed such phenomenal growth during its first fifty years that it wasn't unlike Standard Oil or Carnegie Steel. In fact, that's how Roosevelt saw it. As Carnegie's and Rockefeller's empires had been dismantled, so too was that of the Balls'. In 1939 FDR's administration launched an antitrust lawsuit against Ball and eleven other glass manufacturers, alleging they were monopolizing the glass-making industry. In early 1945 the US District Court for the Northern District of Ohio found Ball in violation of the Sherman Antitrust Act, and the Supreme Court, a few months later, affirmed the decision, which meant the end of the Ball brothers' expansion. The brothers' mustaches had gotten too big.

The ruling made modernizing glass factories pointless; Ball's only choice was to diversify. So Ball diversified, and how. The company got involved in every age: the plastics age, the computer age, the space age. Ball made display monitors, pressure cookers, Christmas ornaments, roofing, nursing bottles, prefab housing, battery shells, and a chemical for preserving vinyl LPs. In the early 1980s, Ball made about 12 billion pennies—or rather, 12 billion copper-plated zinc penny blanks for the US mints in San Francisco, Denver, Philadelphia, and West Point. Ball cranked them out at 22,000 per minute. Having made a few thousand four-cylinder cars, and six WWI tanks, Ball built antennas on F-35 Joint Strike Fighters, and instruments that went to Mars. Ball built irrigation systems in Libya and got into petroleum processing in Singapore. Ball made engraving plates for newspapers. Ball helped fix the blurry vision of the Hubble Space Telescope. And Ball got into cans.

By the late 1980s, Ball's stock price was stuck around $7 per share, and over the next five years, after a small blip, it sunk lower. The company had diversified and expanded so dramatically that sales had increased, while profits had not. In 1993, things turned especially bad: flooding on Midwest vegetable farms and a poor Canadian salmon catch reduced demand for cans. The plastic market further invaded the glass market; only Snapple bucked the trend. Ball closed factories in Oklahoma and California, which meant $58 million lost in tax benefits. Then, because of delays and quality problems in the start-up of a huge new glass furnace in Louisiana, the company lost some old customers. On top of that, changes in accounting practices burdened the company with a $35 million bill. Share prices sunk lower, and earnings that year were below Wall Street estimates every quarter. By the end of the year, Ball reported a loss of $33 million. The next year, dividends dropped below four cents—half of what they'd been. It was time to consolidate and refocus. It was in that climate—annoyed that Wall Street investors refused to see Ball as anything other than jar maker—that Ball spun off its glass jars under a company named Alltrista. Its Nasdaq ticker: JARS. Ball jars are still made by Jarden Corporation, which bought Alltrista and uses the Ball name under license. Ball, in the meantime, had figured out how to crank out cans 250 times faster than it ever produced glass jars.

If you stacked up all of the aluminum beverage cans produced in a year, the stack would be 13.5 million miles long. That's long enough to make a tower that reaches the moon and to have enough cans leftover to make fifty-five more such towers. Of course, since an empty can is only capable of supporting 250 pounds, and each can weighs about a half ounce, you couldn't stack up more than 7,353 cans before their own weight would crush the bottommost can, toppling the whole thing. So, practically, you're limited to building a tower 2,757 feet tall, which is 40 feet taller than Dubai's Burj Khalifa, the world's tallest skyscraper. With all those cans we pump out every year, you could make 20 *million* such towers, which means you'd have to build more than 50,000 of the highest man-made structures ever built every day just to keep up with production.

With the cans produced at Ball's plant in Golden, Colorado, which I visited on the second day of Can School, you could build 816 of those towers daily. Every day except Christmas and Thanksgiving, the plant spits out 6 million cans. That's a semitruckful every twenty-two minutes. The plant—Ball's biggest beverage can plant in North America, employing a few hundred people around the clock—is in an industrial part of town, down the road from a few auto repair shops, across from a paving company. Covering fourteen acres, the plain building sits only five miles from the center of town, where Coors (in partnership with Ball) also makes cans, making Golden the can capital of the world.

Erich Elmer, the plant's tall, spectacled, and scholarly assistant manager, led a small group of us on a tour. From start to finish, it only takes about an hour to make the most-engineered product in the world. Of the twenty steps involved in can making, I was most excited about the twelfth step, where machines spray the internal coatings. But because we'd be passing by many powerful machines, and because a manager once broke all of his toes when a forklift carrying a three-thousand-pound coil of aluminum ran over his foot, Elmer had us put on ear protection with built-in headsets, safety glasses, and bright green vests before going onto the factory floor.

The plant was tritonal. All of the machines were green. All of the moving parts of the machines—the gates and presses—as well as safety marks on the floor, were yellow. The floors were a uniform, shiny gray. It was therefore easy to tell where to go and where not to; where to put your fingers and where not to. But because a line, as it's called, is all about efficiency and speed, much transpired so quickly that it was impossible to see it.

Elmer led us past massive coils of aluminum sheet, about the thickness of a piece of paper. The coils, almost 15 tons and a body-length wide, looked like giant rolls of toilet paper. Unrolled, some are a mile long. Dozens were stacked on the polished concrete. One coil fed into the cupping press, a 150-ton cookie-cutter machine that stamped out sixteen circles at a time and sounded like a locomotive. Not far away, a conveyer belt delivered the cups to the bodymaker, an enormous three-stage ram. Another conveyer belt, like a giant toy train, then took the cups to giant trimmers.

At this point, most of the mechanical work forming the metal into cans was done. From the trimmers, the cups made their way to a six-stage washer, more or less a giant carwash. When the cans emerged, at 300 degrees, they were specular, meaning mirror bright. In that condition, they made their way to a printer, where ink and a clear coat were applied to the outside of each can, and to a bottom coater, where a tiny bit of Teflon-laden clear coat was slapped onto each can's footprint, the better to protect the can while it slid along. Then the cans were ferried through an oven, where they were cured for about a minute at 400 degrees.

At this point I began to get a headache, on account of the overwhelming noise. The noise rivaled the loudest cheering I've ever heard, and the headache felt like jet lag. The combination was overpowering. Even with a headset, I couldn't understand half of what Elmer was saying. Meanwhile, millions of cans zipped along on conveyer belts at such speed that the motion was impossible to detect.

And then there were the IC spray machines. There were seven of them, each capable of spraying 320 cans per minute. Overhead, a conveyer belt delivered cans into seven large funnels, and from these the cans dropped, one at a time, into the glass chambers of the machines. As each can dropped, it was loaded onto an indexed wheel, which rotated, stopped, rotated, stopped. As the wheel rotated, the can spun rapidly at 2,200 revolutions per minute. For an instant, each can was a veritable figure skater doing triple lutzes. At the first position, a high-pressure injector sprayed liquid epoxy down into the can. At the second, another injector sprayed the liquid epoxy at a slight angle, toward the top of the wall. This was to get an even, consistent coating. On the walls of the chambers, atomized spray accumulated like fine snow and formed a white, lard-like goop. An employee was scraping one clean with a metal spatula, much as one scrapes frost from a windshield.

And then, in a fraction of a second, the cans were gone. A conveyer belt whisked them away to another oven, which cured the epoxy by heating it up to 390 degrees, gradually, for two minutes.

Beyond the oven and some fans and the waxing machine and the necker and the flanger, Ball had a few machines and workers examining cans, to be sure they had been coated properly—in other words, thoroughly. Five digital black-and-white cameras examined the coating inside every can, with a strobe. One camera pointed straight down, while four pointed at the neck, where the end would eventually be fastened. A computer assessed the gray scale of every can—at two thousand cans per minute—to determine if the coating was sufficient, which is to say perfect.

Another test employed a light sensor to test for holes in every can. The tiniest pinhole could spell disaster: a leaker or an exploder.

Line workers also collected sample cans, to test the internal coating more precisely. Hourly, they filled these samples with an electrolyte and then measured its current (in milliamps). If the coating wasn't perfect, the electrolyte would come into contact with the aluminum, electrons would start to migrate, and the measurement would not read zero. They call this a metal exposure test. Line workers also tested the coating with two adhesion tests. In one, they measured how much force it took to scratch through the coating. In the other, they scratched the coating, then slapped sticky tape on the wound and pulled it off, and measured how much coating had been removed.

When things go wrong with the IC sprayers, apparently, it's usually obvious. The injectors rarely execute partial sprays; they either fail to spray, or spray too much, and then the coating doesn't cure right. If the can isn't clean or still has lubricant on it, the coating doesn't stick. If the oven settings are off, and the coating heats up too fast, blisters form. Unlucky customers have found, in their sodas, the internal coating floating on the surface. I'd take that over a mouse any day.

Finally, the cans were stacked in imposing pallets. A pallet of 8,169 cans is nine feet tall. They're stacked two, three, four high in a warehouse that, with thirty-foot doors, outdoes all other warehouses except airplane hangars. You could fit a house through its front door.

～～～

The world's first cans, developed by an Englishman in 1810, owed their corrosion resistance to a layer of tin. It didn't hurt that the tin canisters were a fifth of an inch thick, and weighed over a pound. It was one of America's first can makers, a Londoner who sailed to Boston named William Underwood, who is responsible for the shortened word that we know today. While he was sterilizing (pasteurizing) tin canisters of lobster in a giant iron kettle on the beach in Harpswell, Maine, one of his bookkeepers shortened the word *canister*—from the Greek *kanastron*, "a basket of reeds"—to *can*.

The art of can making entailed so much trial and error that early cans weren't reliable. They exploded. They exploded because can makers didn't like sterilizing their cans. Boiling them in water made them rust. That's why some canners afterward slathered their cans with lead paint or lacquers. And still they went bad. Unlucky canners lost 100 percent of their production; lucky ones dumped thousands of errors in the Erie Canal. Some warned dealers to test their cans before selling any. A Wisconsin canner of peas, who quartered in the second floor of his warehouse, lost sleep on account of all the cans of peas exploding below him. Bacteria were to blame, and a young professor of bacteriology at the University of Wisconsin named Harry Russell studied and solved the problem for the pea canner in 1894. Can making had finally turned scientific.

In the meantime, can makers had begun using tin plate coated with enamel, which allowed them to can applesauce, sardines, and tomatoes (without juice). Then something funny happened. A Maryland canner complained about black spots in his cans of corn. He figured the tin plating was no good. Herbert Baker, the chief chemist at the American Can Company's lab, figured him wrong. He showed that "corn black" was iron sulfide, the iron from the can and the sulfide from the corn. He also showed that zinc prevented its formation. Zinc had been in the solder used in old handmade cans. New solder, and new machine-made cans, didn't contain any zinc.

From 1911 to 1922, Baker worked on getting zinc back into cans. First, he looked at the thickness of tin, and the purity of the steel, but these were dead ends. Then he impregnated parchment paper with zinc oxide, and lined cans with it, which was effective but inefficient. Galvanized ends worked, too, but imparted upon foods a zinc flavor. Finally, Dr. G. S. Bohart, chemist at the National Canners Association, put zinc oxide in enamel.

Like modern epoxy coatings, enamels separated the package from the product. Can makers had learned the hard way that a plain tin-lined can was fine for pastas, peaches, pears, and pineapples, but would bleach strawberries, cherries, and beets so thoroughly that customers wouldn't return to buy more. Peas, any kid knew, were mild, but beans, which contained sulfur, would turn blue, then black. Bohart's C enamel (as opposed to the standard R enamel) worked because the zinc oxide reacted with the hydrogen sulfide before it had a chance to react with the can. C enamel allowed can makers to put all kinds of heretofore forbidden—that is, corrosive—products in cans. Can making companies quickly figured out how to put ham, dog food, and orange juice in cans. Aggressive foods, such as sauerkraut, pickles, and jalapenos, would eventually demand even thicker coatings.

Still, nobody could figure out how to can beer. Tin-coated steel cans turned beer cloudy, and ruined its taste. Iron was even worse: just one part per million of iron in beer ruined its flavor. This was because of iron's interaction with water: it tears water molecules apart, releasing oxygen that changes the taste of the beer, and corrodes the can. C enamel wasn't cutting it. Brewer's pitch, as sticky as tar, seemed a good coating candidate, except that it didn't survive pasteurization. The solution was two different coats of enamel—one made by Union Carbide, and the other made by a small company that became Valspar. In 1935, after three years of effort, American produced the world's first beer can. Gottfried Krueger Brewing, of Newark, New Jersey, was American's first customer. Pabst got into the game six months later. By the end of the year, can makers had sold more than two hundred million steel cans to twenty-three brewers.

By the time Bill Coors, an engineer, borrowed a quarter million dollars and set out to manufacture aluminum cans, in 1954, he could turn to epoxy rather than enamel to protect the metal. He could also turn to the work of an English scientist, Denis Dickinson, who in 1943 had made a stab at quantifying the corrosivity of the stuff in a can. He called his measure a corrosivity index. To compute it, he took a two-inch strip of a can, boiled it in hydrochloric acid for two minutes, and measured how much metal had been lost. Usually it was somewhere between 100 milligrams and 300 milligrams. Then he took the same strip of metal, put it in a food or beverage for three days, at 77 degrees, and measured how much more metal had been lost. A product's corrosivity index was the ratio between metal lost to food and metal lost to acid. Most products came out below 1. Fruits came out anywhere from 2 to 4. Anything above 6, Dickinson figured, was abnormally corrosive. Yet he remained unhappy with his measure. "It is unfortunate that this, probably the most important diagnostic factor," he wrote, "is still so imperfectly understood." He'd have loved Ed Laperle and been fascinated by Mountain Dew.

Whatever the constitution of his energy drink, Bob doesn't just put it in one of Ball's cans and close it. He must deal with the headspace: the 5 milliliter air bubble at the top. He doesn't want oxygen in there, because oxygen trapped there will dissolve in the beverage and corrode the can. If what's on the inside of the can gets on the outside, terrible things happen. In a warehouse as in a hospital, just one untreated infection can be enough to infect every patient in the building.

Bob deals with this threat by flooding the headspace with carbon dioxide or nitrogen before seaming the can shut. If he doesn't, he voids the warranty on the can, exposing himself to all kinds of liability. Because this final step happens at a beverage plant, not at Ball's can-making plant, Ball assists. The guy who oversees is Dave Scheuerman. A dozen reps work for him, running around North America, enlightening customers on "double seam theory," teaching them how to read good cans, recommending pressurizers and optical oxygen sensors, helping them properly put their product in cans. Scheuerman has been working at Ball for thirty years and has seen a lot of good cans perish in the line of duty. He tells their tales somberly, soberly, and slowly. I met him on the third day of Can School, and of all the engineers I met, he was my favorite. A trained food biologist, he seemed more like a general in the Ministry of Can Defense. With an undershirt poking out beneath a blue shirt and jacket, he looked not unlike a tubbier Robert Redford. He knows that even with all the technique in the world, there's still room for error. "It's a talent," Scheuerman said. "An art form you have to learn."

The simplest mistake the filler at a beverage manufacturer can make is one of overcompensation. A filler may crank up the nitrogen or carbon dioxide and overpressurize his cans. Other beverage manufacturers play it safe by overfilling their cans, selling 12.2 or more ounces of beer for the price of 12 ounces. This is not a rookie mistake. Scheuerman has done the math, evaluated engineering tolerances, and calculated with assurance that a manufacturer who decreases his fill range one twentieth of an ounce—from 12.1 to 12.05 ounces—will save 174 cases, or 4,176 cans for every million cans he produces. Fewer of his cans will explode. It's technically 0.4 percent of his inventory, constituting a few thousand dollars' worth of product that he was giving away for free by overfilling, and enough product to fill 17 more cans. It pays for itself: no destroyed products, no threats to other perfectly good products, no wasted time, no complaints, no lawsuits. "If they have too many duds," Scheuerman said, "I just tell 'em, back off on your fills."

Sometimes duds manifest themselves on account of small changes in the way a particular beverage is made. A new fertilizer turns out to be corrosive, an ink turns out to release lead, a new coating turns out to produce traces of benzene, or a lemon-lime extract, imported from Brazil, turns out to have been boiled in a copper kettle. Some copper ends up in solution, and once in the can, together with acid, corrodes it galvanically (recall the Statue of Liberty). The can plates out, and the plating removes more of the internal coating, and the bare aluminum only exacerbates the effect. Copper can also come from the water supply, the water that is added to soda syrup. "You don't even need a microscope," Scheuerman said, describing the symptoms of copper contamination in cans. "You see black dots all over the can. You see a reddish copper tinge."

Overgassing, overfilling, and unwittingly contaminating cans are of such concern to Scheuerman because two out of three leaking product complaints now come from warehouse problems. This is the dreaded outside-in corrosion. Scheuerman laid out the scenario: a primary leaker leads to secondary leakers and in not much time the distributor of Bob's Energy Drink is facing an unfolding nightmare. "You get a leak in a can at the top of the case," Scheuerman said. "The can leaks, drips down, wicks into adjacent cases. You end up with a Christmas tree pallet." He paused. "I have seen entire warehouses, with a million cases, that were a total write-off." On the podium at Can School, he said this seriously, like it might pain attendees to imagine. Rocking back and forth on his feet—heels, toes, heels, toes—he said calmly, "This is not a salvage operation."

Hot warehouses are particularly terrifying to Scheuerman. In such places, the alloy used to make the ends "relaxes," or loses 7 percent of its strength. So warehouses in Alaska are safe. Warehouses in Alabama are risky. A 115-degree shelf is a problem; a 150-degree trunk—like Lynda Ryan's—is definitely a problem. Hence the disproportionate number of failures from the South. Worse, shrink-wrapped pallets, in hot and humid warehouses, tend to collect and trap condensation at the can's most vulnerable spot, the top. Technically, there's no coating on the top of the score mark around the aperture, so if water collects there, you get top-down corrosion, and leakers, and then more corrosion problems. (Food cans sometimes rust via the labels, which absorb moisture.) So Scheuerman recommends using shrink-wrap with slits in it. Otherwise, he said, truck drivers delivering cans will hear pops in their trucks. "The sad thing is," Scheuerman said, "they always ask, 'Is this safe to send to the customer?' I don't know. More may pop next week." He says a fan is okay, but what distributors really ought to do, ironically, is keep the product warm—warmer than the dew point.

Scheuerman learned most of this fifteen years ago, when Ball started printing 1-800 numbers on cans, and people called with problems. "We figured, of all the places our cans go, supermarkets, gas stations, and convenience stores, that the last two were where the problems are." He was wrong. Most consumer complaints were coming from supermarkets. Next in line was vending machines. He wondered: how can the claims be so high? It turned out they'd find a leaker, wipe the others off, and put them back. "They were spreading a cancer," Scheuerman said. "In vending machines," he said, "if they act quick, they get away with it. If not, they gotta replace the whole machine and clean it." He said all of this with a tone that would not be inappropriate at a wake. His face remained straight.

"When a customer picks up a can and it's dented, banged up, corroded, or leaky, they usually don't say, 'Boy, Safeway beat the heck out of this,' or, 'the Freightway Truck Line abused this.' They look at it and they say, 'I can't believe these people put this on the market, on the shelf.'" Ball doesn't like it when that happens. So Scheuerman gives customers corrosion bulletins and displays. And he has a road-ready presentation. He can do it anywhere. All he needs to know is the number of people he's presenting to, and what language to do it in. Calmly, steadily, Scheuerman balances defense with offense. From lessons learned in the field, he suggests modifications that increase efficiency and reduce the chances that a canner will be exposed to a lawsuit from a can that explodes and takes someone's eye out.

Still, cans fail. Ball implores customers to notify them ASAP if they have any complaints of leakers, and compiles the responses in a report. They want to perform a "root cause analysis" on all field failures they can get their hands on. During Can School, Laperle showed me where they do this. It's called "the morgue," and it was a small room near the flavor lab where thirteen cans, dead of various causes, lay exposed on a black countertop.

It's Laperle's task to determine if the fault of these field failures lies with Ball or not. Sometimes, a canner or distributor or consumer sends a can back because of an unidentifiable blob. This is not a problem. About a third of the lab contains machinery devoted to solving such a mystery. There's a Fourier Transform Infrared Spectroscope, which checks the composition of the blob against a database of 100,000 chemicals, and a gas chromatograph, which checks a database of even more. The blob could be lubricant, or ink, or any other number of manufacturing chemicals, or it could be something the filler goofed on, like something from a truck, or a pallet, or it could be bird poop, or part of a mouse, or alfalfa pollen, which has turned beer the color of Mountain Dew.

To be sure it's not an impurity in the metal, there's an electron microscope in a room next door. With this, employees like Michelle Atwood, a young Midwestern biochemist, examine the crystal structure of the metal in the can, looking for globs of iron, for example. When I poked in the room, Atwood zoomed in on the tiny back-to-back *R*s incised into the tab of a Rockstar Energy Drink can. She zoomed from 100x, to 500x, to 1000x, until the backbones of the *R*s looked jagged, like thorny spines of a rose bush.

Often the fault lies with customers who unintentionally discover the weakest point of the can. These customers tend to be old, and they tend to winter in Florida. In March, these snowbirds buy cases of soda, put them in the closets of their trailers, and drive north. In October, when they head south again, they open the closet door to a swarm of fruit flies. One or more of the cans have burst. This happens regularly, every fall: perfectly made, perfectly filled, perfectly warehoused cans fail.

The cans fail because the ends—which are manufactured on a separate line at the Golden plant and get shipped to canners in brown paper bags, like huge packages of Ritz crackers—are more susceptible to corrosion than any other part of the can. To make a can openable, the aperture must be scored such that kid fingers and elderly fingers can lift up on the tab and tear the aperture. The score line is only $\frac{1}{1000}$ of an inch thick, and technically it's not coated. The die that forms the score is so sharp, and so forceful, that the coating right at the score line is removed. To be sure the scores are just right, they're camera and pressure tested at the Golden plant. "Just right" is an understatement. "Give or take two or three millionths of an inch," Elmer, the plant's assistant manager, once told me while pointing to a can, "and this can won't open." Elmer grew up in Missouri, making parts at his father's machine shop. The parts were for the aerospace industry. Cans, he explained, are manufactured with much tighter tolerances than aerospace parts—ends particularly. Many in the industry like to say, he told me, "The can is just a pedestal on which the crown sits." Marvel of marvels: not perfect-tasting beer but the top of the can.

But that perfect end remains susceptible to corrosion. That's why the snowbird move is so detrimental. The only thing touching the inside of the end ought to be harmless carbon dioxide (or nitrogen). Leave a can sideways for six months, and it'll leak. "I've often wondered," Scheuerman told me, "why they don't write 'Store upright' on the cans."

It took two years of research before I met anyone who had experienced an exploding can. As it happened, the man who told me of that unlikely event wasn't a snowbird but Jamil Baghdachi, of Ypsilanti, Michigan. In the summer of 2006, Baghdachi was on his way from Louisville, Kentucky, to the airport. It was rush hour, and just as he drove his Volvo under an overpass, a soda can on the passenger seat exploded, spraying all over his face. To Baghdachi, it was not a disaster. He is the director of the Coatings Research Institute at Eastern Michigan University. He has consulted on can coatings and has dozens of coatings patents. As he explained, "It was a lack of a complete coating."

⌒

Two hundred years ago, it took an hour to cut, assemble, fill, and solder shut the first cans. And then, because there was no other way to open them, people attacked their cans with knives, bayonets, hammers, and chisels. They smashed them open with rocks or fired rifles at them. For fifty years, even though their contents had often spoiled, cans seemed like marvelous containers. Can openers were invented, and cans seemed even better. By the turn of the twentieth century, with can-making machines spitting out one hundred cans a minute and pasteurizing techniques down pat, the can seemed incredible as ever. A generation later, so did Ball jars, but you could get beer in a can. Surely this metal container could improve no further.

A generation later, this container was made of aluminum. Not long after, manufacturers learned to crank out a thousand of them a minute. They learned to make them thinner and lighter, openable not with a hammer or a church key but with a riveted tab and scored panel. They learned to crank out two thousand of them a minute. They learned to line them with such durable plastics that beverages of nearly unimaginable corrosiveness could be put inside. They painted them in thermochromic inks, lined them with blue coatings, devised resealable screw tops. Having made trillions of these things, Ball now makes cans accurate to 50 millionths of an inch, and aims for errors to be so rare that they are six standard deviations from the mean. The only thing Ball can't make is clear aluminum, so that customers might see the water they intend to put inside. Cans have improved so much that engineers at Ball—self-declared perfectionists—debate whether or not they can improve any further.

On the second day of Can School, Mary Chopyak, who has spent twenty-one years as a materials engineer at Ball, pointed to the asymptotically shaped curve of can weights over the last fifty years, and announced that we're at the tail end of improvements. In the last twenty-five years, cans have only gotten one-hundredth of a pound lighter. She said that there are only minuscule improvements left to be made in the can. Sandy Deweese, an engineer who works on can ends, said, with a slight drawl, that nothing else can change. He said that can engineering has reached its limit. Yet Dave Wrenshall, a technical advisor at Ball—tough, mustachioed, somehow reminiscent of Robert De Niro—said it's hogwash; that we're not near the end of optimization. He kept his feet squarely planted, as if there were magnets in his shoes. "People have said that for the last twenty years," he said. He didn't mention foreign objects or explosions, or epoxy internal coatings.

⌒

When I first heard about Can School, I called a public relations employee at Ball named John Saalwachter. Saying that he was excited for me to attend, Saalwachter put me in touch with his boss, Ball's director of corporate relations, Scott McCarty. This was a couple of months ahead of time. Then I did some preparation and asked a few too many questions. Emailing Ball's general counsel, I asked about the various liability concerns relating to foreign objects and explosions. This probably raised some flags. I also inquired about the internal coatings, and this certainly raised more. Two weeks before Can School, McCarty called me. In no uncertain terms, he told me I couldn't come. He said that Can School wasn't for journalists; that no journalist had ever attended. That it wouldn't be fair to those in the industry.

It wouldn't be fair to the industry, I came to conclude, because the internal coatings used to prevent aluminum cans from rusting from the inside out and adulterating their products are as secret and controversial as fracking fluid. The formulas used by the major coatings manufacturers—PPG Industries, Valspar, and Akzo Nobel—are proprietary; the details vague. I called each of the major manufacturers; none returned my calls. In legal documents, details are carefully marked confidential and redacted. In patents, formulas are left vague—citing, for example, a component added in a range anywhere from 0.1 percent to 10 percent. Even the US Food and Drug Administration (FDA) must, as one chemist there put it, "pry stuff out of" the manufacturers. Often the formula is unpatented and therefore preserved as a trade secret.

This much is clear: the epoxies must be affordable, sprayable, curable, strong, flexible, and sticky—such that not much beyond sulfuric acid or the potent solvent methyl ethyl ketone (MEK) will remove them. Creating such a material takes a cross-linking resin, curing catalysts, and some additives to give it color or clarity, lubrication, antioxidative properties, flow, stability, plasticity, and a smooth surface. The resin is usually epoxy, but it may also be vinyl, acrylic, polyester, or oleoresin, and could even be styrene, polyethylene, or polypropylene, or a natural drying oil derived from beechnut, linseed, or soybeans. The mixture also requires either a solvent, so that the epoxy can cure when baked, or a photo-initiator, so that the epoxy can cure when exposed briefly to ultraviolet (UV) light. The cross-linking agent of choice for the most tenacious epoxy coating is bisphenol-A, or BPA. BPA is the primary ingredient in such coatings because it makes the plastic plastic.

In addition to thickness, which is determined by the corrosivity of the product, beverage can coatings engineers pay attention to how the can will be treated. Will it be pasteurized? Stored in a hot place? Shipped on a vibrating train through someplace hotter or colder or more or less humid? They tweak the application or select a different coating as required. For beer, engineers might use a coating, patented by Valspar, that contains cyclodextrin, a donut-shaped carbohydrate that, via molecular inclusion, traps bad-tasting molecules. Food can coating engineers have greater concerns. Coatings for tomatoes must be stain resistant, those for fish must resist sulfur, and those for fruits and pickles must resist acids. There's one coating for tomatoes, one for beans, one for potatoes, and another for corn, peas, fish, and shrimp. Chocolate, especially sensitive to adulteration, requires its own coating. Those for meats must contain a lubricious wax, called a meat release agent, so that the meat slides right out. Fruits and vegetables including beets, currants, and plums, which contain the red pigment anthocyanin, are some of the most corrosive. The top of the list belongs to rhubarb. It, alone among foods, requires three layers of lacquer, and even with that much protection, rhubarb still boasts a shorter shelf-life than its peers. All told, there are over fifteen thousand coatings, and though most of them serve inside food cans, many of them perform their work inside beverage containers. Beverage cans in the United States demand about twenty million gallons of epoxy coatings every year, for about one hundred billion cans. According to coatings specialists, roughly 80 percent of that epoxy is BPA. A tiny bit of it, then, ends up in us—reason, perhaps, for the secrecy.

Biologically speaking, hormones are rare, and potent. The system that produces, stores, and secretes them—the endocrine system—controls hair growth, reproduction, cognitive performance, injury response, excretion, sensory perception, cell division, and metabolic rate. Endocrine organs— including the thyroid, pituitary, and adrenal glands—produce particular molecules that fit into particular receptors on cells, unleashing a chain of biochemical events. Hormonal changes in infinitesimal quantities cause dramatic changes, including diabetes and hermaphrodites. Endocrine disruptors, including molecules that mimic the hormone estrogen—called estrogenic chemicals, or xenoestrogens—get jammed in the cells so that the real molecules can't get in there and do what they should. Others fit perfectly, triggering events the body didn't intend to initiate.

Clues of synthetic chemicals having such effects, detailed in Rachel Carson's 1962 book *Silent Spring*, had been piling up since the 1950s. Beginning in the late seventies, wildlife biologists around the Great Lakes, primarily studying fish and birds, found that compounds were altering the cells, bodies and behavior of the animals in novel ways. They found masculinized female fish, feminized male fish, intersex fish, and birds that refused to raise their young. In a landmark 1993 study, biologists Theo Colborn and Frederick vom Saal described the range of this "endocrine disruption," a term coined only two years earlier.

BPA's role as an endocrine disruptor wasn't recognized until 1998. That's when Pat Hunt, a geneticist at Case Western Reserve University, in Cleveland, noticed something strange about an experiment she was running on some mice. Forty percent of her control mice—the ostensibly normal, healthy ones—were producing abnormal eggs. "We checked everything," she told the writer Florence Williams. After weeks of ruling out potential culprits, including the air in the lab, Hunt noticed smears and scratches on the animals' plastic cages. It turned out someone had used an acidic floor cleaner, rather than a mild detergent, to clean the cages, and it was degrading them. Something was leaching into the mice's feeding tubes and interfering with the mice.

The something was BPA, an artificial estrogen first studied in the 1930s to prevent miscarriages. It didn't work for that—it has quite the opposite effect—but the double-hexagon-shaped molecule was soon put to work making polycarbonate plastics shatterproof, which made them the future, and abundant. It took a half century, but scientists—finally studying chronic toxins rather than acute toxins—caught up with plastic. A 2011 paper titled "Most Plastic Products Release Estrogenic Chemicals," published by the National Institutes of Health's National Institute of Environmental Health Sciences, sums up the current understanding. In it, researchers describe detecting estrogenic activity in over five hundred commercially available plastics, including many advertised as BPA free. They report that manufacturing processes—such as pasteurization—convert nonestrogenic chemicals into estrogenic chemicals, and they note that sunlight, microwave radiation, and machine dish washing accelerate the leaching of estrogenic chemicals. They also mention that the most effective way to release estrogenic chemicals is by using a mixture of polar and nonpolar solvents (in their case, salt water and ethanol), which pretty much describes Bob's Energy Drink.

According to Hunt, who doses mice with amounts of BPA proportional to their body weight, she sees abnormal behavior or cells in the mother, her offspring, and her offspring's offspring. One dose damages three generations. This is because BPA shows up in cells in the breast—the body organ most sensitive to the effects of estrogen—which means that its effects get passed to the next generation at the most fragile, sensitive time. Timing turns out to be a major factor. In fact, it turns out that exposure to BPA may be comparable to exposure to DDT, the carcinogenic insecticide about which Carson wrote. Girls exposed to DDT before puberty, researchers found, have five times the cancer rate of girls exposed to the same amount afterward.

Hunt, Colborn, vom Saal, and many other researchers around the world have found that BPA can cause early puberty, obesity, and miscarriages, lower sperm counts, and increase rates of cancers of the breast, prostate, ovaries, and testicles. All these troubles were observed in rodents, many of which were exposed while in their mothers' wombs. In other rat experiments, prenatal exposure to low doses of BPA caused lesions in mammary glands. Other studies have confirmed that BPA activates the estrogen receptors on breast cells, and can cause cancer cells to replicate in a dish. BPA has been shown to cause normal breast cells to act like cancer cells. It's also been shown to be just as powerful as diethylstilbestrol, or DES, a very strong synthetic estrogen discouraged since 1971 in pregnant women because of its association with rare, awful reproductive cancers.

Hence, perhaps, the retraction of my invitation.

After McCarty told me to steer clear, I was devastated. I spent a morning searching for people who'd been to Can School (it's amazing what people put on their résumés) and the rest of the day trying to get in touch with them. I emailed a friend at a local brewery and asked if any of her colleagues were going. I got nowhere. The day after McCarty's call, I received an email from Clif Reichard. The subject was Ball Beverage Can School.

I figured it was something from a lawyer. The first sentence of the email said, "We have you registered to attend our Beverage Can School." I read it a few times before continuing. It went on, telling me where and when to show up. It said that lunch was included, that dress was business casual, and that open-toed shoes were not allowed for the tour of the plant. It had Reichard's cell phone number. And it thanked me, ahead of time, for coming. I didn't know what had happened. Maybe the higher-ups at Ball had changed their minds. Maybe they goofed up. I didn't care. I was in. Twelve days later, I drove down to Broomfield.

Ball's Broomfield headquarters, halfway between Denver and Boulder, is situated on an awesome piece of property, offering a full hundred-mile panorama of the Front Range of the Rocky Mountains. From north to south, Longs Peak, Eldorado Canyon, Mount Evans, and Pikes Peak are visible. Driving in, I was nervous. I noticed two video cameras on the south side of the driveway, just after a sign that said PRIVATE DRIVE. I wondered: Would lawyers pounce on me? Would security? Might I get escorted out? Or arrested?

From the parking lot, I walked up a path, under some trees, and through two sets of dark glass doors. The lobby within was a sunny atrium, with a dozen leather chairs in the middle. On the right, there was a flat-screen TV with Ball's stock price ticking across the bottom. It was $36.15. On the left, there was a security desk, and past it, an enlarged sepia print of the five Ball brothers looking all business, their mustaches lifesize and resplendent. I headed for the security desk.

Sweating, I tried to play it cool. I said I was there to attend Can School. The attendant asked for my name and began scrolling through a stack of ID cards. What was my name again? He couldn't find my card. He asked for a driver's license and had me enter my name on a sign-in sheet. Then he handed me a temporary badge, which said Escort Required. Being badge-less seemed advantageous. I could be anybody, or nobody. Still, I stayed low, not flaunting my presence. The attendant pointed me down the hall, to a conference room. He warned me not to deviate without an escort.

I walked in, and sat in the second row, near the right, by myself. Nervous, I sat there as if departure was imminent. I kept trying not to look back over my shoulder. I tried to listen to the hushed small talk around me. I was more paranoid than I am proud to admit. Ten minutes later, a security guard brought me a real name tag, and a nameplate for my seat. Below my name, it said Scripps—the name of the journalism fellowship I was on. It seemed vague enough, almost like a brand of soda. Five minutes later, Clif Reichard came over, said sorry, and said he'd give 'em hell for taking me off the registration list. I said, really, it's no big deal, I live just up the road. Really, it's no problem.

During the initial presentations, I didn't ask many questions, because I wanted to stay under the radar. I later gave in and chatted up two employees from the Can Manufacturing Institute. Megan Daum, CMI's sustainability director, was apprehensive. Her body language made it clear that she wanted out. Joseph Pouliot, CMI's vice president of public affairs, was chatty, and smart. When I pressed him about CMI's unremarkable motto that cans are "infinitely recyclable," he admitted that glass, too, fit the bill.

By the end of the day, I was worried again that lawyers awaited. Paranoia was getting to me. Would they demand my notes? My recorder? The thumb drive and documents they'd provided? Not wanting to write that in my notes, I wrote in Spanish, scrawlier than normal: "*puso memoria en pantalones, y uso español en mi papel,*" which was my tired way of noting that I'd put the thumb drive down my pants and used Spanish to conceal that fact.

A good coating is hard to find, and as much effort goes into creating and applying one as goes into any other aspect of manufacturing cans. This was not taught at Can School.

Making an epoxy coating starts with a petroleum refiner like Exxon-Mobil, which produces huge quantities of benzene. Dow Chemical Company or Momentive Specialty Chemicals converts the benzene to bisphenol-A, and combines it, about 4:1, with epichlorohydrin. Dow calls this stuff D.E.R. 331, or Dow Epoxy Resin 331, and Momentive calls it Epon 829. These blends, among dozens, are suitable foundations for coatings on everything from cans to cars and bridges. Next, a chemical company such as Cytec Industries or Sartomer or Rahn buys the epoxy resin, adds 5 percent acrylate to make an acrylated epoxy called something like Genomer 2255, and sells it to one of the big coatings companies. To render the coating just right for Bob's Energy Drink, the major coatings companies then add small amounts of pigments, surfactants, adhesion promoters, corrosion inhibitors, light stabilizers, toners, extenders, thixotropic agents, dispersing agents, wetting agents, dyes, and catalysts—which are often made by yet another chemical company. With their own IC spray machines, the coatings companies line cans, and make sure that the coating doesn't feather when the tab is lifted, craze when dented, or blush in the presence of certain solutions. The finished product costs about $25 per gallon.

The dime-sized nozzles through which the coatings squirt are manufactured and studied and chosen with similar attention. They're made by the Nordson Corporation, in Amherst, Ohio. Nordson, which boasts recent growth not unlike Ball's, makes an immense variety of "dispensing equipment," including more than 1,500 airless nozzles just for coatings. Engineers at Nordson, minding turbulence and vortexes, study how the coatings atomize as they are forced out of a tungsten carbide orifice a few thousandths of an inch wide. The atomized molecules emerge in a flat spray, such that a uniform film thickness distribution may be achieved. For every nozzle Nordson engineers produce, they also produce a map of its spray pattern. In this way, they also ensure the cardiac health of can makers.

The spray machines—a half ton and $2 million each—are designed within equally precise parameters. Ball's spray machines are made by Stolle Machinery, in suburban Denver. Tom Beebe, a salesman who's been working on cans since 1970, showed me one of the machines as it was being assembled by a mechanic. This alone was a feat, requiring consent to take no photographs, since Stolle is as secretive about its machines as Valspar and Nordson are of their coatings and nozzles. Beebe wouldn't discuss coatings, and two IC spray machine engineers in the office were equally reticent. The first one said, "I have no comment," and insisted I get approval from higher-ups before talking. The second walked by without comment.

Ball, which annually spends about $200 million on coatings, stores the stuff at its Golden plant in two ten-thousand-gallon stainless steel tanks, eight feet in diameter and two stories tall. During Can School, they were not part of the tour.

The second day of Can School started ominously, with dark clouds over the Rockies. I sipped from a can of tomato juice—a can-making achievement of its own, since, according to one chemist, "tomato sauce is probably the nastiest stuff you're gonna come across"—and listened to a range of Ball employees: the VP of sales, the manager of manufacturing engineering, the director of graphics services. I heard Mary Chopyak say that she spent two years working on BPA-free coatings. I heard Dan Vorlage, Ball's head of innovation, describe being stymied by his work on BPA-free coatings. Then, at eleven thirty, Paul DiLucchio, Ball's training and development manager, tapped my shoulder and whispered that someone wanted to see me in the hallway.

I grabbed my recorder and notes, and prepared never to step back into the room. I assumed the worst was in store. Walking out, I could feel my heart beating. In the hallway, Scott McCarty was waiting. He had his cell phone to his ear. I figured that meant others were on the way.

McCarty got straight to the point. He asked what I was doing there. I told him that I thought he or someone at Ball had changed his mind and invited me. I mentioned the email from Reichard. McCarty repeated himself, and asked if I hadn't understood him when he'd called. I said I had, but then Reichard invited me. McCarty looked annoyed. I told him that I had announced who I was at the security desk, and that if Ball had meant to keep me out, the company had had the opportunity.

McCarty changed tactics, told me he thought rust was a silly subject to write about, and asked how cans were related. I told him that I was writing about can manufacturing and all of the associated processes devised to prevent rust. McCarty said he still didn't think my book was a good idea or that anybody would want to read about cans. I thought, "That's why I'm me and you're you," but I said, "Let me worry about that." I also told him I'd already talked to Laperle and had taken a tour of the corrosion lab. McCarty dismissed Laperle. "He's pretty much retired," he said. "He doesn't even work for us anymore." "Really?" I said. "Last time I checked, he was the director of your corrosion lab." (He still is.) The conversation went around in circles for fifteen minutes. Finally, he yielded, begrudgingly. Since I'd already attended half of Can School, he said, I might as well stay for the rest.

From then on, though, I was persona non grata.

⁓

Positioned squarely between the industry and the thirsty consumer is the FDA's Center for Food Safety and Nutrition. Federal food-safety regulations, established a generation ago and amended since then, require anything in or touching foods and beverages to meet certain standards. So, after a coating manufacturer designs a coating that satisfies its rheological and organoleptic dreams, a hurdle remains before Bob can employ it to defend cans from his new energy drink.

The approval process is a world unto its own, hidden within the FDA and a handful of private labs. The labs—including Avomeen Analytical Services, Intertek, SGS, and Eurofins—run simulations to see if coatings are fully cured and stable and to determine the toxicology of the migrants that inevitably emerge. The FDA just reviews the data or asks for more. In shorthand, labs call these tests 21CFR tests, after the particular section of the code of federal regulations that requires them, but, really, they're full migration studies with new food contact substances. The results are reported in FCNs, or food contact notifications. They take a few months to produce and cost about $100,000. Hundreds of pages long and incredibly technical, they are not a pleasure to read.

Here's how the process works: a coatings company sends Avomeen a box containing a pint of a new coating and ten stainless steel plates covered with it. An Avomeen chemist takes two-inch squares of the coated metal and places them in a hot simulant—either vinegar, ethanol, or olive oil. According to an FDA chemist named Mike Adams, these simulants accurately mimic package-product interactions. There the metal pieces sit (in triplicate) for two hours, a day, four days, ten days. In each case, he evaporates the simulant, weighs the residue, and computes its concentration.

Then Avomeen chemists examine the migrant or migrants they've found. From a new coating, they might find only one analyte or a half dozen. If it's a chemical that's been studied, they can look up its properties. If it's something unknown, they study it, and how they study it is determined by its concentration. But here's the key: if the analyte shows up in concentrations lower than 0.5 ppb (parts per billion), the FDA considers it good. Below that threshold, Avomeen need not conduct mutagenicity or carcinogenicity tests. Below that level, the FDA requires no multigenerational feeding studies.

Behind a hefty regulatory cover letter drafted by lawyers, Avomeen then submits its report. The FDA, which normally reviews about a hundred FCNs a year (at least half of which are for new coatings), then has four months to raise any objections. Since 2000, about 90 percent of FCNs have passed. They're accessible to the public online. The details—like the chemicals that emerge in minute quantities—are not.

The Can Manufacturers Institute is, not surprisingly, addicted to discussing the benefits of aluminum. The organization infinitely recycles its line about aluminum being infinitely recyclable. The CMI also extols every other virtue of the can: it's cheap, stackable, eminently shippable, and safe—all of which was evident by the conclusion of Can School. The organization announces that cans prevent light and UV penetration better than glass or plastic bottles, that cans cool down faster than glass or plastic, that they keep out oxygen better. It points out that food-borne bacteria kill 5,000 people a year and hospitalize another 350,000, averaging $1,850 annually per US citizen. Affirming that no such illness has resulted from a can for thirty years, it praises the greatness of the can. Remember, it insists, that a fifth of all American dinners include something from a can. When BPA concerns are raised, they cite studies they maintain confirm that BPA levels in cans are safe.

The North American Metal Packaging Alliance similarly dismisses the importance of BPA. In a 2011 press release, NAMPA urged policy makers and media to "take heed of a decisive analysis by independent toxicologists that concludes bisphenol-A (BPA) poses no risk to human health." NAMPA's chairman, John Rost, has called coatings critical and safe, and encouraged decision-making restraint to keep any legislator from acting with haste. Rost, a trained chemist, is also a lobbyist. BPA coatings, he has said, provide superior performance, without even a "marginal health risk," and have been reviewed by health agencies. That, and there's no readily available alternative.

The American Chemistry Council, too, seeks to allay worries over BPA. At the websites Bisphenol-A (www.bisphenol-a.org) and Facts About BPA (www.factsaboutbpa.org), which were registered by the ACC, the group debunks nine BPA myths, mostly by pointing to the lack of a "sound scientific basis."

Ball employees haven't developed a coherent strategy. Some attempt a mild deception, by calling the epoxies "organic coatings." Of course, aldrin and dioxin are organic, too—and toxic in minute quantities. Others employ a few more syllables, calling the epoxies "water-based polymers." Mention BPA, and everyone at Ball seems to get tense. Scott Brendecke, the corrosion engineer, stammered so after I mentioned BPA that he lost his train of thought. Paul DiLucchio said BPA is perfectly fine and that everybody refuses to understand as much. Another employee shrugged suggestively, raising one eyebrow at the mention of BPA. All he said was, *"Hey, intent."* With that—wink wink, nudge nudge—he walked away. To that end, I heard one employee call Ball a "practical provider of solutions," and another say he just provided options.

Can makers argue that modern society offers plenty of exposure to BPA outside of cans, and that it's been deemed safe; that the quantity of BPA in each can is minuscule; that even less migrates into beverages; that the quantity detected in humans is even smaller ("extremely small"); and that, regardless, any absorbed BPA gets expelled daily in urine. "There is a danger of over-reaction to issues relating to migration if the available data is not put into the context of the actual low levels to which consumers are exposed," writes the can consultant Bev Page. They say that the relationship between BPA and health effects is associative at best, and furthermore, that studies on mice aren't relevant in people, and that the results of many such studies are not reproducible. Finally, they say that major regulatory bodies in Japan, Australia, New Zealand, Canada, and the United States agree that current levels of BPA exposure are safe, and that one, the European Food Safety Authority, recently recommended increasing the tolerable daily intake of BPA by a factor of five.

"Agree" and "safe," though, are at odds with the opinion of the US Health and Human Services Department. HHS has said that parents should do all they can to limit BPA exposure in their infants. "Concern over potential harm from BPA is highest for young children," the agency warns, "because their bodies are early in development and have immature systems for detoxifying chemicals." The National Toxicology Program takes a similar stance: "The NTP has some concern for effects on the brain, behavior, and prostate gland in fetuses, infants, and children at current human exposures to bisphenol A." It ranks concern on a five-step scale, from negligible concern, to minimal concern, to some concern, to regular concern, to serious concern. It has some. The American Medical Association feels the same way. Even the FDA recently declared: "Results of recent studies using novel approaches and different endpoints describe BPA effects in laboratory animals at very low doses corresponding to some estimated human exposures. Many of these new studies evaluated developmental or behavioral effects that are not typically assessed in standardized tests."

Frederick vom Saal, who has been studying hormones as long as Scheuerman has been studying cans, has found that BPA is as potent as DES, able to act far below the FDA's threshold, below 1 part per *trillion* (ppt). In 2004 the CDC found that of 2,517 people six years and older, 93 percent of them had BPA in their urine. A 2012 study in Ontario found that workers in food canning (not can making) had twice the risk of breast cancer than the general population, fivefold if they were premenopausal. A 2008 study in Shanghai found a threefold risk, as did a 2000 study in British Columbia. The authors wrote, "It is plausible that they were exposed to BPA from can linings."

While Americans debated uncertainties, other countries made decisions. Canada added BPA to the list of toxic chemicals under the Canadian Environmental Protection Act. France voted to continue with a ban on BPA-based polycarbonate bottles. Denmark voted to continue with a ban on BPA in food packaging meant for children under age three. Japan has nearly eliminated BPA can coatings. The United Nations Food and Agriculture Organization and the World Health Organization convened a four-day joint meeting on the subject. Yet a commonly held view in America is that best expressed by Governor Paul LePage of Maine. "The only thing that I've heard," he told the *Bangor Daily News*, "is if you take a plastic bottle and put it in the microwave and you heat it up, it gives off a chemical similar to estrogen. So the worst case is some women may have little beards."

⁓

At the conclusion of the third day of Beverage Can School, DiLucchio—the man who had tapped me on the shoulder—thanked the attendees, expressed his gratitude, and told everyone about the certificates awaiting us on the table beside the door.

I collected my stuff, chatted up a few more people, and took a look. There were a few dozen certificates on the table, in alphabetical order. They said:

In recognition of your contribution to the Ball Corporation Beverage Can School, on April 12–14, 2011, in Broomfield, Colorado

Joe Shmoe
Is hereby inducted into the
ORDER OF THE CAN MAKER

after completing 16 hours of can school training, and is entitled to all the rights and benefits thereof, including the right to high quality beverage cans and responsive customer service from Ball Corporation.

John A. Hayes,
President and Chief Executive Officer, Ball Corporation

Michael Hranicka,
President, Metal Beverage Packaging Americas

Those with last names at the end of the alphabet—Vuolo, Wang, Wonson, Zeund, Zivkovic—were at the bottom-right corner of the table. I looked again, but couldn't find Waldman. Everybody got a certificate except for me.

A month after Beverage Can School, I returned to Ball headquarters for Food Can School. I showed up fifteen minutes early; this time, I was kicked out before the program started. McCarty told me it was "over capacity," though there were a half dozen empty seats. When I observed, afterward, over email, "It sure *seems* like you're trying to hide something," he responded three minutes later: "I'll take that as an attempt at humor. Feel free to send me any questions." He never responded to my questions.

⁓

A half century ago, in *Silent Spring*, Rachel Carson wrote, "For the first time in the history of the world, every human being is now subjected to contact with dangerous chemicals, from the moment of conception until death." Carson described chemicals accumulating in the sex organs of mammals. She described low sperm counts in pilots who sprayed pesticides. She described the terrifyingly abrupt deaths of chemical handlers who goofed up and accidentally touched products they didn't know not to. She called some of the chemicals "elixirs of death."

When Rachel Carson wrote *Silent Spring*, she noted that one out of four Americans would get cancer in their lifetimes. Now the rate is almost twice that. Since she died in 1964, we've detonated 1,439 more nuclear bombs around the world, about half of them in the southwestern United States. We've banned DDT, but we've also introduced BPA into so many things that it's easier to list the products that don't contain it. *Chemical World News*, in a tone today echoed by the CMI, NAMPA, and the American Chemistry Council, called *Silent Spring* "Science fiction, to be read in the same way that the TV program *The Twilight Zone* is to be watched." I thought the book was far more frightening than any TV show, old or new. Carson wrote, "If we are going to live so intimately with these chemicals—eating and drinking them, taking them into the very marrow of our bones—we had better know something about their nature and their power." She didn't even know many of the compounds had the ability to alter human hormone systems, or that we would produce products containing them billions of times over, marketing them to children, selling them in vending machines in schools.

As Florence Williams writes in *Breasts: A Natural and Unnatural History*, "Our pursuit of the good life—controlling reproduction, smoking, drinking, reading fine literature while lazing about indoors—also contributes to our plight . . . We've overloaded a fragile biochemical machine. We're a living mismatch between our wiring and our desires." Williams knows. She had her urine tested for BPA. It had 5.14 parts per billion, in the upper 25 percent of Americans like her, but still four hundred times lower than what the US Environmental Protection Agency considers a safe dose. She changed her diet for a few days, and got her urine tested again. She got it down to 0.759 ppb. She's a mother of two. She had her six-year-old daughter's urine checked too. It was 0.786 ppb.

Patrick Rose, a lawyer who sued Ball unsuccessfully in a 1999 exploding can case, is more concerned about BPA than explosions. "I won't drink out of a can," he told me. "I won't let them near my kid. I'm kinda surprised this hasn't trickled down into the popular culture." He said he thinks the huge increase in breast cancer incidences is related to BPA. "We're ruining women's health," he said, "and it's totally preventable."

Jamil Baghdachi, the can coatings consultant who runs the Coatings Research Institute, is similarly fearful. "It scares the hell out of me," he said. "I'd rather have a glass. What if it's not cured? What if some of the chemicals exude out?" He said, "I've seen the chemistry; I see it on a daily basis . . . I know what can go wrong, how it can go wrong." He calls cans "suspicious" and doesn't buy canned food. He told me he'd like to write an op-ed—as Florence Williams did—in the *New York Times*. "The less you know, the better your life is. The more you know, the more you worry . . . There's no equation for it. It's a perception. I've got that perception because of my awareness."

Frederick vom Saal, the respected biologist, also won't buy canned foods or beverages, and won't allow polycarbonate plastics in his home. In a 2010 interview with Elizabeth Kolbert, in the Yale University online magazine *Environment 360*, he recalled an act straight out of the industry playbook. When he studied a dose of BPA 25,000 times lower than anybody had studied, in 1996, and found developmental harm, Dow suggested that he not publish his results. When he did, BPA manufacturers called and threatened him. So did the chemical industry. He said that toxicologists are off by anywhere from one to eight orders of magnitude. For regulatory agencies, he reserves greater criticism, calling them "locked into procedures decades out of date," unable to acknowledge, let alone perform, modern science. He's published studies on endocrine disruptors in two dozen journals, including *Nature*, the *Journal of the American Medical Association* (*JAMA*), and *Proceedings of the National Academy of Sciences* (*PNAS*). He called the system fossilized, a lie, and a fraud. "This is the highest volume endocrine-disrupting chemical in commerce," he told Kolbert. "We don't know what products it's in. We know that in animals, it causes extensive harm. There are now a whole series of human studies finding exactly the same relationship between the presence of Bisphenol A and the kind of harm shown in animals. That scares me. I don't think that's alarmist. This is a chemical about which we know more than any other chemical with the exception of dioxin. Right now, it is the most studied chemical in the world. NIH has $30 million of ongoing studies of this chemical. Do you think that federal officials in Europe, the United States, Canada, and Japan, would all have this as the highest priority chemical to study, if there were only a few alarmists saying it was a problem?"

Mike Adams, the FDA chemist, is not reluctant at all to drink out of cans. "I know how we evaluate these things," he said. "We've got, like, the most nervous toxicologists on the planet. They're very careful about their decisions. I don't have any hesitation." Of the simulations, he said, "we are one hundred percent sure that this is not gonna be a problem." Of BPA concerns, he called it "paranoia." He told me that the testing labs are "really, really pushing levels of analytical capabilities."

Ball employees are not hypocrites: they eat and drink out of their cans. Some employees believe that cans will keep getting better; that this end-of-the-line talk is nonsense. Others say there's no more room for improvement: that cans are already over engineered, and manufactured with minuscule failure rates. That today's can—cheap and utilitarian and wonderful—is the best possible mousetrap. But it's not perfect. Because most see BPA as a business concern rather than a health concern, they refuse to admit that BPA-free cans would be an improvement.

Oddly enough, Ball's already done it. The company makes BPA-free cans for a small Michigan company called Eden Foods. Four varieties of beans and four types of chili are available in BPA-free cans. According to Sue Potter, the marketing director and wife of the company president, they tried to put tomatoes in the cans, but they were too acidic. Actually, plain tomatoes were okay, but tomatoes with garlic, onion, and basil weren't. Ball said they'd last for only six months. Potter told me that the BPA-free cans, made with an oleoresinous enamel—that is, a natural, oily coating—cost 2.5 cents more than standard cans. "I think everybody should do it," she said. "I don't get it."

Everybody dances around what to call the can's internal corrosion inhibitor. The FDA calls it a resinous and polymeric coating. Ball calls it an organic coating, or water-based polymer. The EPA calls it a chemical pollutant. Health researchers call it an endocrine disruptor, and a chronic toxin. Everybody's dancing because most of us are addicted to cans. We're no more capable of giving them up than we are of giving up beer.

The Breast Cancer Fund, in San Francisco, has urged its members to contact food makers—Del Monte, General Mills, and ConAgra in-cluded—and demand BPA-free cans. Legislators in half the states in the country have introduced bills banning BPA products, but most of the bills died as so many do, in commerce committees. America produces millions of gallons of BPA a year, for profits exceeding $6 billion. Few elected officials are going to stick their necks between that and the American Chemistry Council, the Grocery Manufacturers Association, and the US Chamber of Commerce.

I asked Jamil Baghdachi what he thought of amending the government warning that appears on alcoholic beverages, such that it said, "according to the Surgeon General, women should not drink alcoholic or canned bever-ages during pregnancy, because of the risk of birth defects." He said, "A lot of people nowadays are reading labels. Will it help? It will help. It's a step for-ward. It's like global warming. The sooner we begin the discussion, by labels, by books, by articles, the better it is." But he emphasized that he personally wouldn't advise the change, because the industry would shun him. He said an authority like the Surgeon General would have to make the change.

⌒

Long after Can School, I stumbled across two homages to the can as the perfect container. Both precede the can-making industry's reliance on BPA-based epoxies and the publication of *Silent Spring*. Both predate the industry's recent turn toward secrecy, lobbying, denying, delaying, and de-ceiving. In that sense, they seem naive, for there's no hint of maneuvering between cost and health, or between quantity and quality.

Metal cans, filled with beer or Mountain Dew or Bob's Energy Drink, yearn to rust, but to these men, the stakes were only unsightly black dots or an unpleasant off-taste. Before the era of endocrine disruption, these men took pride in having briefly stopped entropy. Where most of us never noticed the crown atop the pedestal, these two men did.

The first homage was delivered by Colonel William Grove of the Quartermaster Corps of the US Army. In February 1918, at the annual convention of the National Canners Association, in Boston, he recited this poem:

We can march without shoes;
We can fight without guns;
We can fly without wings
To flap over the Huns.
We can sing without bands,
Parade without banners,
But no modern army
Can eat without canners.

The second was delivered by William Stolk, CEO of the American Can Company, creator of the first beer can. In New York, on April 21, 1960, he said,

There is a fashion today, the world around, of judging material progress in terms of the big and the spectacular. Underdeveloped countries, for instance, vie with each other as to who shall have the biggest steel plants, the tallest hydro dams, and the most important refineries. And then, in the same way, progress is sometimes measured in terms of missiles, rockets, and space ships. We at American Can Company have never entered either of these spectacular races—for size or speed. Most of our one hundred twenty–odd plants are comparatively small. Our products are opened today and thrown away tomorrow. Our most cherished achievements have been unknown and all but invisible to the consumer. Yet they rank with the telephone, the automobile, and the electric light in the revolutionary effects they have had on modern living. They have lightened the work of millions of housewives. They have opened up for farmers vast new markets for quickly perishable foods. They have helped eliminate such dietary diseases as scurvy and pellagra. They have made possible the supermarket, the almost clerkless food store, and the ser-vantless household. In fact, they have helped make possible our largest cities and smallest towns, and without them our whole population would be quite differently distributed.

I want to feel like those men. I want to be a can evangelist. But I'm torn, because I'd also like to raise a kid someday, and I'd like that kid not to be exposed to a potent endocrine disruptor for the sake of convenience. I'd like to have more faith in industry and government, and feel like I did on the second day of Can School, before I got pulled aside, when I was drinking coffee from a paper cup, marveling at the only thing there not in a can.

5

INDIANA JANE

The rustiest place in America is not open to the public. Patrolled by private security guards and town police, the site is enclosed by a tall chain-link fence, which bears these warnings:

PRIVATE PROPERTY
NO TRESPASSING
VIOLATORS WILL BE PROSECUTED

NOTICE:
THIS AREA UNDER SURVEILLANCE

DANGER
KEEP OUT
PELIGRO

The place is the Bethlehem Steel Works, in Bethlehem, Pennsylvania. Once the world's second largest steel producer, it has been rusting since the middle of the Civil War, when iron was first made there. Until the mid-1970s, when dust filters arrived, rust from "the Steel" coated the surrounding city too. It settled on windshields and windowsills, and prevented residents from hanging laundry out to dry. Old steelworkers, correlating more rust with more steel production, swear they could tell from the thickness of the rust how big their paychecks would be. In 1995, with the American steel industry in shambles, the paychecks stopped, and the last blast furnace shut down. Since then, the place has done nothing but rust. Now, from the air, the abandoned complex looks like a decrepit brown castle in an otherwise green city.

One woman is exceptionally familiar with the place. Her name is Alyssha Eve Csük. (Her last name rhymes with book.) The granddaughter of a steelworker, she is a photographer. She photographs rust. She is, as far as I know, the only person who makes a living finding beauty in rust. As such, I joined her at the motherlode—which she calls her playground—on a snowy late-November day, to see how she does it.

I met Csük at a downtown Bethlehem coffee shop. In her late thirties, she wore jeans and a tan sweater. Long, streaked blond hair fell below her shoulders. Of medium height, she seemed half stoic and half distracted. On her advice, I bought a muffin and a bagel, for a later snack, and stowed them in a brown paper bag in my jacket pocket. Then we walked across the street, to her studio, where she answered some emails and geared up for a day at the Steel. While she got ready, I marveled at her prints, so much more vivid in person than on her website. Many, from her collections *Abstract Portraits of Steel* and *Industrial Steel* and *the Yards* and *Slate Abstracts*, were leaning against a wall next to her front door. Some were stacked on top of a wide cabinet, and many more were inside sliding drawers. A stack of small prints lay on a kitchen table. Above Csük's desk hung a spooky, cerulean print that looked simultaneously in and out of focus. A twenty-year-old splotchy-brown cat named Sweet Pea followed me around.

Aside from three small Rodin sketches, almost everything in the studio hinted at Csük's photography career. On a bookcase, I saw photo books on Henri Cartier-Bresson, Marina Abramović, and Mary Ellen Mark. Next to a blank to-do list, I saw this, assembled from a magnetic poetry set: "you are a wild universe goddess & your art belongs in a world of a thousand dreams." It wasn't far from the truth. Her work, which has been featured in photo magazines and the *New York Times*, hangs in galleries and private homes and corporation lobbies. Corrosion, as Csük sees it, isn't brown and dreary, nor does it suggest age and decay. It's alive and glowing, much more wonderful and exciting than silver. Though a few of her prints seem merely Pollockian, most are far more compelling, evoking the skin of some wild animal, or the patina on Western sandstone, or the aurora borealis, or the lick of flames. When she zooms in on metal, she captures speckled reds, lumpy yellow waves, green crests, serrated blues, orange slashes. One print looks like a Japanese watercolor, another like Japanese calligraphy. My favorite, which I nearly drooled on, suggested a thin cascade of the purest blue water over the darkest Yosemite granite.

Eventually Csük emerged in black North Face ski pants, a gray turtle-neck sweater, and a long, black, hooded down jacket. She wore gray hiking shoes and soon put on red half-finger biking gloves with leather palms. She packed two camera bags: a black backpack and a green canvas shoulder bag. She also grabbed a carbon-fiber tripod. Then we hopped in her SUV.

Making art requires bending rules, and the same goes for Csük and her rust art. Technically, she has permission to enter the fenced-off steelworks—property now owned by the Bethlehem Sands Casino Resort—as long as she stays on the ground level. When this does not appeal to her, which is often, she sneaks in. With me, she snuck in.

Csük drove a mile south, just over the Lehigh River, and parked near the New Street Bridge. Under the bridge, we crossed five sets of railroad tracks, then ascended the grassy levee separating the tracks from the river, and took a right. I held the tripod close, to conceal it. A half mile ahead, five two-hundred-foot blast furnaces loomed. Csük walked toward them with purpose. She'd contemplated walking out of sight, on the rocks along the river, but the snow made that route treacherous. On the grassy levee, the snow just soaked our feet. Halfway to the Steel, a white pickup truck approached from behind on the gravel road beside the tracks. As the bearded driver—presumably a railroad employee—passed us, he waved. Having no choice, we waved back. Later, I wondered if he had called us in.

Five minutes afterward, in the shadow of the Steel, a few obstacles stood in our path. The first was a train, stacked two high with containers, parked on the middle track. Fortuitously, it blocked us from sight. Csük looked both ways, and then slid down the slippery levee, and climbed up and over it. I followed close behind. She looked both ways again, and jogged over to the second obstacle, the chain-link fence. When she realized that we had left footprints in the snow, she stepped back and tried to brush them away, which only made them worse. From there, we walked along the fence in gravelly spots, so as not to leave footprints. I followed her a bit farther—past the no-trespassing signs—and then, just before noon, we climbed up and over. Alyssha went first, and then me.

Over the next five hours, I watched Csük wander around a mazelike industrial complex of greater entropic value than a sub-Saharan market, calmly and boldly, without a map, in search of aesthetic minutiae that most people miss entirely. To reach good vantage points, she scampered atop a large pipe, thirty feet up, and along a giant crane, even higher. She set up her tripod seven times and took sixty-nine exposures. Only once all afternoon did she seem nervous, and not on account of heights. First, though, she hurried through a courtyard overgrown with shrubs and vines and littered with glass shards and old buckets. Massive brown tanks loomed above. She hurried because she was not comfortable out in the open, where she was visible. She made her way to blast furnace D, her favorite. Then she climbed a few steep flights of rusty stairs. Immediately, on the streaked wall of an enormous gas stove, she saw something appealing. A layer of metal pipes had been removed from the stove and tossed into a huge pile on the ground, and now a new rusty surface was visible.

She said, "There's something beautiful here. I don't know if it'll fit my format. I'll have to see it with my camera. This is probably just gonna be a sketch." She opened the tripod and placed it on a metal grate. She put her camera—a Canon EOS 1D Mark IV, with a 35-millimeter lens—on the tripod. Behind it, she half-squatted, with her left knee down, her right knee up, and her right elbow balanced on her right knee. The pose was very Rodin. She looked through the camera. She moved the tripod back two feet. She said, "I can't use this lens. Too much distortion." She put the 35 millimeter in her right jacket pocket, and put on a 24–105 millimeter. She zoomed to 100 millimeters and raised the tripod a hair. "Just like I thought, this really doesn't fit my format. There's the potential for something. The image is just a square, but I'm trying to make it fit. Lemme try moving back a little bit."

Csük might spend fifteen to forty-five minutes fiddling with a composition. In this case, she could tell it wasn't worth it. Before she packed up, she looked at me and asked, "Did you hear that?" I told her I thought it was the sound of a motorcycle somewhere in town. She said, "Sometimes, things fall here." She told me later that thirty- or forty-pound objects—heavy enough to guarantee death—rain down regularly.

Csük climbed another flight of stairs and walked to a spot where more light struck the stove. She walked slowly, with her head tilted a bit to the left. She said, "I wish we had more of this going on, like a whole brigade of this. Over here is beautiful. I gotta shoot this." I looked, saw no formation—brigade or platoon or even a mere patrol. She continued, "This was stuff I never got to see before, because it was all covered up. And this'll weather more, 'cause it's all exposed." Positioning the tripod back a few feet, she hunched on both knees. Her head remained askew, soothsayer-like. Then she moved the tripod a few inches. She looked through the viewfinder and moved the camera a few more inches. She looked again, and moved the camera a bit to the right. Then a few inches back. Then up a hair. Then to the right a hair. Then up a bit. Finally, with a shutter-release cable on a three-foot cord, held in her right hand, she took a shot. She leaned back, resting her head on a rusty railing, and said, "It's kinda nice."

On the opposite side of the towering stove, at a spot facing south, Csük set up again. She said, "There might be some vibration, so I'll take a bunch." The vibration was from a train going by eighty feet below. Even without trains rolling by, she might take dozens of exposures and still not get a shot. She might return five times to the same scene, at all hours of the day and night, in all varieties of weather, and still not get it. This time, though, she had a feeling. "There's something really beautiful about this image," she said. She took exposures from 0.8 seconds to 2 seconds. Then she moved the tripod a few feet to the left. "This isn't nearly as interesting," she said. She took one shot. She moved the tripod a few more feet left. Four shots. Satisfied, she stood up and looked out over the railing, as if to take it all in. She noticed some green moss she hadn't seen before and followed her nose in its direction. Checking out the metal, she walked out of sight.

Just then, I heard a noise. My pulse rose, and I froze.

Csük has had plenty of run-ins and close calls at the Steel. She's nearly bumped into all kinds of vagrants and wanderers, and always spotted them before they spotted her. Once, up on a crane with only one way down, she heard voices in the room below her. She stood still for a half hour until the men left. On a different occasion, she nearly crossed paths with a lunatic from West Chester, who shortly thereafter was arrested and found to be in possession of many guns. Two hundred feet up, she's nearly stepped through a staircase missing four rungs. While poking around with another photographer in 2005, she suffered her closest call. In blast furnace E, she encountered a handful of people, and the pair ran to hide in a dark corner of blast furnace D. On her way through the cavernous room, she fell through a rectangular hole where casts of molten iron were drained from brick channels into railroad cars below. According to the other photographer, one moment Csük was there, and the next she was mostly gone. "Had it not been for her backpack and camera and tripod, she'd have fallen down to the bottom," he told me. "It was far enough to kill her." It was, more precisely, two stories down to the ground. The photographer grabbed Csük by the armpits and pulled her out. She'd smashed an expensive Linhof lens and scraped her left leg, but suffered no other injuries. As a result, she never rushes inside the Steel. To this day, the other photographer, who has returned to the Steel many times with Csük, calls her Indiana Jane.

Csük reappeared and said, "There's something really magical from here to here. The color is all there." Not wanting to seem anxious, I didn't mention the noise. Then she told me to follow her. She led the way down dark staircases, around corners—none of it familiar. I was sure it was a new route, until, outside the building, I saw our footprints leading in. I was impressed. Csük told me that she's always felt her way around, yielding to whatever draw she senses, as if blind. She just trusts her instincts, and she's bombarded with things that catch her eye. Of shooting rust, Csük admitted, "It's funny bringing life from something so lifeless." She seemed, all of a sudden, like a field biologist or a mountaineer.

She walked up a corrugated ramp, under a wire fence, and along an elevated trestle on a warped platform. At one point, she stopped, looked both ways before proceeding, and told me to follow close behind her, because we'd be visible. Then she ducked under the wire fence and walked down a different ramp, into another building.

From a ledge in this building, she looked west, pointed to a giant exhaust valve inside a cage of beams, and said, "Oh wow. I wish I could photograph that." The valve was at least sixty feet up, overhanging a tangly courtyard. "I really love what's happening there now. With the red and the black forming, it's very intriguing. I can't get it. You get used to that here." She couldn't get it because her tripod needed to be stationary, on terra firma. Rigging wouldn't work, and was too risky. She said, "That is so jungle looking. So tribal." She turned around, ready to keep exploring, and then turned back. "Now, why can't that be accessible?" Then, lower, almost directed at the steelyard itself, she said, "You're killing me."

Csük made her way higher, to a perch with an expansive view. She looked out at snowy roofs, admiring the way snow collected on different surfaces, angles, features. Off of one spot, the wet snow avalanched, leaving stripes. On another, the snow caught drips, and appeared speckled. On a curved section, the snow faded gradually away. Csük noticed a wire hanging down she hadn't seen before. She said, "I'm just taking it all in." As she later told me, she has no problem standing somewhere and looking and looking and looking. She's patient. She's not a fast talker, or a fast walker, or a fast eater, or a fast typer, or a fast anything, except maybe driver. She rarely drinks coffee, and never before shooting. I had mistaken Csük's refusal to rush earlier at the coffee shop: it's not stoicism, but a focus so intense it slows her down. I watched Csük as she watched, and saw a long stare take hold of her. Her eyes got glassy. She kneeled. All of a sudden, she sprung back to planet Earth. "I don't know why that car is stopping," she said. She told me to stay where I was, at the edge of the window frame, just out of sight. In the meantime, she froze. "They're not looking up here, are they?"

Csük took her first photo class for the hell of it, after a few unsatisfying years working as a medical transcriptionist for a handful of vascular surgeons in a medical office next to Lehigh Valley Hospital. Realizing she needed more, that the nine-to-five was not for her, she considered going into psychology or design. She weighed the decision for weeks and decided to pursue the latter at Northampton Community College. Her studies began with Drawing 1. She hadn't drawn since childhood, and even then, hadn't really drawn. Her parents had let her doodle on a wall, and while her brother drew people and houses and trees, all she drew was abstract shapes. Her mother had attended art school, had won awards for her paintings, and never pushed Csük toward art. If anything, Csük grew up intimidated by the artistic talent in her family. Since then, she'd been scared of drawing and art. At Northampton, she learned how to draw, and discovered that she had what one teacher called artistic dyslexia. She drew light stuff dark, and dark stuff light. She drew negatives. Realizing this, she took a photography course. She liked it so much, they gave her the keys to the darkroom.

In her twenties, after a boyfriend introduced her to museums and galleries, the opera and Philip Glass, she decided she wanted to further her photography studies. She applied to the Rochester Institute of Technology, and in November 2003 was accepted for the term starting the following year. At the end of that month, on a warm breezy day, she was on her way up Perry Street when low light coming through the ruins of the Steel caught her eye. She rode her bike toward the Steel, hopping off at a fence. She grabbed the fence, captivated, and resolved to return immediately with her first digital camera. Two days later, not knowing any better, she walked the route we had, without scouting for cops first. She felt only anticipation. A cop, seeing her, blared from his car, "You have now entered private property." She turned around and returned to the bridge, until she was out of sight. Then she scampered down to the rocks along the Lehigh River and followed it a half mile to the Steel. She peeked over the levee. Then she ran across the train tracks, and into the rustiest place in America. Her heart was pounding. A train passed just as she gained entry, hissing as she stepped in.

Her curiosity became a pursuit, and before long it seemed like a worthy photography project she could complete before she ran off to Rochester. She figured it was a thirty-day project. In the ensuing eight months, she spent forty-six days at the Steel. By the start of the semester, she didn't want to go to photography school. She wanted to keep shooting rust.

At RIT, Csük began studying photojournalism, but it wasn't for her. She wanted to learn as much technical craft as possible, so she switched to advertising—which meant a lot of time in a studio. She thinks this has benefited her. When she graduated, a professor asked her what she planned to do. She told the professor that she wanted to be a fine art photographer, and the professor just laughed. He said, "Do you know how hard that is?" Says Csük: "I had no encouragement. It was just cold, like boot camp."

Now, nine years later, almost to the day, she—the country's preeminent rust photographer—was still at the Steel. It was her 377th day there.

The car moved on, and so did Csük. She descended convoluted flights of stairs, walked back up the ramp, ducked under the wire, looked both ways, and proceeded a hundred feet down the elevated track. She stopped, took a left, ducked under the wire, and walked down a different ramp.

By then, I had begun to look around the way Csük does. I pointed out some compelling swirly drips on an overhead pipe. Csük had seen it many times and tried to photograph it on at least a half dozen occasions. "It's mesmerizing," she said. Then: "I wish—I can't get up higher. It's torture. I'm constantly tortured with photos that I can't get." To capture the swirls, she said, she'd need to be there early, on a snowy day, so that the reflecting light bounced up without lens flare. There was determination in her words. The pipe wasn't going anywhere.

Halfway down the ramp, Csük caught an iridescent glow fifteen feet up on a boiler. "It's beautiful, but I only shoot straight on," she explained. Her head lowered, she scanned at shoulder height and saw something. "Wow, it's beautiful. If you just shift your angle, all these colors come out." She extended the tripod to four feet. "It's kind of interesting. I'm like a hawk. Who would see that? I don't know if I can capture it." She positioned her camera, took a few frames, and said, "A lot of people would say, 'I got it.' I don't feel like I got it." Then: "I'm never done with it. I just keep coming back."

She led onward, stepping over rusty things, onto rusty grates, through rusty walls and door frames. She leaned against railings and beams that seemed like they might give or break. Over crackly ground, littered with clanky debris, she somehow walked quietly. Through a spring-loaded gate, she made her way, seeing that it didn't clack shut. When she sneezed—just once—she did so quietly. The space seemed like the inside of a drum: taut, loud, dark, and echoey. She walked back to ground level, and then up some other stairs, to a ledge, and said, "I like what's happening over there, with the water and the reds, but there's not a good composition." She walked by and then looked back. "I just love those areas of red over there."

Quickly, because they were exposed, she climbed up two flights of stairs and stopped on a platform covered in cracked rust an inch deep. She was hunting for a better angle. Up there, pointing at a wall of orange and brown, she said, "It's amazing how the colors change. On another day, that'd be blue—a glowing blue." From the platform, I heard a male voice below. I told Csük. Immediately, she backed into a dark room and stood there, silent. She looked at me, and said, "Fuck. It's Joe."

Earlier that morning, Csük had told me about Joe. Joe Koch. For twenty-six years, he'd worked at the Steel as a safety officer, and was now employed there as a security guard. A handful of times, he'd stopped her at the gate on Founder's Way and revoked her pass. The last time this happened, he tried to dissuade her by delivering a forty-five-minute lecture on copper thieves, who, he said, would probably kill and bury her, at which point Koch would have to get the cadaver dogs out. Csük had looked at me and said seriously, "He's probably here today."

Csük was first officially granted access to the Steel in 2004 by a local developer. By the terms of an indemnification agreement, she was supposed to stick to the safety of the street level and stay out of dangerous areas like the blast furnaces. When the Sands acquired the land, her access was transferred. Nevertheless, security guards—many of whom had worked at the Steel most of their lives—stopped her regularly, revoking her pass. They wanted her to cough up $30 an hour to be guided around the site, saying that it was not a safe place for a young woman. Csük, though, had a friend in a Sands executive in Las Vegas, who saw that her pass was reinstated indefinitely. In time, one of the casino's designers got in touch. He wanted to see what she'd been up to all those years.

When Csük showed the images she'd made, the man was impressed. So were the new owners. Many of the security guards—the lifers—were floored. "They just never knew that rust could be beautiful," Csük said. "They say, 'I just never saw rust that way before.'"

Recognizing the obsessed visionary in their presence, the casino hired Csük to document the redevelopment of the steelworks. Redevelopment meant destruction of the place that had enthralled her for years. From 2007 through 2009, all but one of the buildings surrounding the blast furnaces—the mills, the foundries, the forges, the tool shop, the machine shop, the basic oxygen furnace, open hearth furnace, the electric furnace, the Bessemer converter, the sales office—were gutted, destroyed, and leveled. Parking lots were paved in their place. Unobtrusive landscaping was installed. Around the blast furnaces—the only sacred thing remaining, according to Csük—a fence was erected.

Csük documented this massive transformation, thinking a book would come out of it. She says it was like watching a slow death. Many of these images, in her collection *Industrial Steel*, seem reverential, as if the steelyard were an iconic peak or pristine canyon. The environs and their contents are clear: walls, rooms, cranes, coils of wire. They're shot like landscapes, at dawn, at dusk, under moonlight, in fog, under a blanket of snow. Csük says she spent hours just watching how the light changed on the yards. She says she photographed the place as people photographed the West.

Documenting the demolition of the place where she'd become a photographer was just as difficult to reconcile. It forced her to focus on more than just steel and rust, and branch out—into slate and scrap yards and trees. She did this so that her spirit wouldn't die. Then the economy tanked, and the Sands put the book idea on the backburner.

Now that the blast furnaces are all that remain, any further damage is traumatic. Trespassers have vandalized parts of it. Copper thieves have stolen bits and pieces of it. Set producers on the movie *Transformers 2* have transformed part of it. Death, as the poet Wallace Stevens wrote, is the mother of beauty, but only to a point. Hours before, when she climbed to the fourth floor of blast furnace D and noticed that metal pipes had been removed from the stove, she'd looked down, and said, "Oh my gosh, look at that pile. That's the guts being ripped out. It's sad." She wants the Steel to suffer a natural death, not an accelerated, assisted, man-made one.

In fact, one of the first things Csük had said to me, in her car, was that if she had a time machine, she'd go back ten years, to the time when the yards had been abandoned but were still open, authentic, raw, and undisturbed. At the time, I didn't get what she meant. The place seemed all of those things to me. Now, she said, the only way to recapture the magical and mysterious feeling of that era was to sneak in. Nighttime exploration, she said, "provides the greatest opportunity for an unadulterated experience." Later she told me that if she won the lottery, she'd buy the steelyard and preserve it.

For two long minutes, we stood still as stones in the dark room. I wondered if the bearded railroad employee in the pickup truck had called us in. If Csük was nervous, she didn't show it. She poked out into the light, the rust crunching under her feet loudly. She looked around. Then she climbed another flight of stairs, where she found a vividly green and red pipe tee to photograph.

"Right now, the colors are just amazing," she said. "It's perfect. I mean, look at the colors. Everything's alive. I've been here before, but it's never quite looked like this." She set up the tripod five feet high, took a few shots, and then got excited. She said, "Oh, I know where we gotta go." Before she went anywhere else, she stood before her tripod, on her tiptoes. Her head leaned left. Steam rose from her breath. The shutter-release cable dangled, swinging back and forth. She thought about her shot. Then she moved the tripod back a few feet, as far as she could, up against a railing. Behind it loomed a five-story drop. The sound of dripping came from all around.

"I'm constantly distracted," she said. "I'm shooting this, but—this is nice. I'm pretty happy right here. I think I got a reward today." By a reward, Csük meant a shot that she loved and could sell. Of the 28,093 rust shots she'd taken before today, 113 were rewards. On this day alone, she got 3.

Depending on their size, Csük's images sell for $800 to $3,200. She sold somewhere between one hundred and two hundred prints in 2012. She sold one, forty-six inches by ninety-six inches and printed on metal, for $30,000. Most are printed on Hahnemühle Museum Etching paper, with archival pigment prints. When I first held one, at her studio, I was almost convinced the image was 3-D, on account of its richness. The detail in the image was so fine that I felt a hunger and had to fight the urge to reach into the print. I admit to telling Csük that I hoped to sell a lot of books so that I could afford one.

"This makes you think of a kite of color, of shapes," she said. I had not thought about a kite. She took a dozen frames and then packed up. Before continuing, she explained her avoidance of the obvious and her desire to look for layers beyond beauty. She said, "A lot of people look at my stuff, and they know it's rust, but they never think it's rust. It can't be too literal."

Csük has been told that one of her abstracts looks like a heart valve. She thinks it looks like an alien planet. Another, she thinks, resembles an elephant. One suggests a mountain range. One suggests *The Old Man and the Sea*. Another suggests a vase. She has been told that her rust photos evoke: a Navajo bird beak, a snow leopard, a field of red poppies, a forest, leaves in snow, a nebula, an amoeba, an abstract nude drawing, the World Trade towers. I'll admit to claiming that one image looked like Einstein. She calls some "primordial," "futuristic," "prehistoric," or "like the fabric of space time"—whatever that looks like. She recognizes a draw to places that feel otherworldly: cement mills, caves, Death Valley. She dreams of shooting rust on a NASA launch pad. But she has done her best work in the town where she grew up, a mile from her home. She uses words like *ethereal, magical, emergent, spirited*. In all of our conversations, she rarely swore. She cites Carl Jung's stream of consciousness, and a philosophy of being, thinking, and seeing described in a book called *Wabi-Sabi*. But for her slight New Jersey accent, she would blend in easily in Boulder or San Francisco.

Csük had gotten excited about a long, cavernous space beneath the elevated track, because in the winter the space harbors huge icicles, and she led the way down to it. Before the final staircase, she warned, "You gotta be careful. People can see you here." I poked in hesitantly and saw some giant gears, but no icicles. Eager to remain concealed, I climbed out quickly. On the way up, I noticed that Csük was using her tripod as a hiking staff.

Like an old sage, she walked casually through a dark room with black paint peeling off a maroon and yellow patch of metal. "This at one time was cobalt blue," she said. "Cobalt blue. It's just amazing how it's changed over the years." She walked around a furnace, and behind massive slag cars—now cauldrons full of green, slushy water. She jumped down a three-foot ledge, and then proceeded into a courtyard, where she looked up at the side of a building. "The whole panel was blue last time I photographed it. Now it's black. We should just try it anyway." A flock of geese flew overhead, following the river. For a second, their squawking sounded like voices.

All of a sudden, the Bethlehem Steel Works seemed like a wonder of the world, a historical artifact as impressive as a pyramid. A few hundred feet away stood blast furnace A, the oldest standing blast furnace of its kind in America. Soon after it was built, in 1914, Bethlehem Steel produced twenty-five thousand shells a day and became known as "America's Krupps." Sixteen years earlier, Bethlehem Steel was the place where Frederick Taylor had worked out how to make high-speed tool steels that could cut three times faster than anything else, which captured the attention of Harry Brearley, the father of stainless steel. Of Bethlehem, its president in the early twentieth century, Charles M. Schwab, used to say he wasn't in business to make steel but to make money. Bethlehem made plenty of money, but it also made bank vaults, battleships, rail ties, and the enormous 140,000-pound axle at the center of Ferris's famous wheel. The company built the USS *Lexington*, America's second aircraft carrier. That beam captured in the iconic 1932 black-and-white photo, with eleven workers sitting on it, eating a carefree lunch eight hundred feet above New York City: that's Bethlehem Steel steel.

Numerous are the elegiacs for Bethlehem. The authors of *Forging America: The Story of Bethlehem Steel* call the Steel "silent," "shuttered," "stark," and "empty." The author of *Crisis in Bethlehem: Big Steel's Struggle to Survive* calls the Steel a decaying, abandoned wasteland, lamenting that pigeon droppings cover idled rolling tables. When I asked Csük about this mourning, she said the backdrop only made coming upon her abstracts that much more delightful. While the authors and historians focused only on the obvious, Csük got to dance between two worlds. "While many may look at these sites as brownfields littered with abandoned buildings and humps of rusting metal," she explained later, "I find in contrast an emerald city of jewels amidst a dark and mysterious place."

The *Times* seemed to agree. In the eight-photo spread titled "Seeing Beauty in the Rust Belt" that appeared on Sunday, May 15, 2011, the author wrote, "During the decline of the American steel industry, Bethlehem Steel's properties have suffered. The result may be bleak, but it's not boring."

Csük wandered on and set up her tripod in front of some metal that she said reminded her of trees in a forest. I did not see trees. I saw drips. I watched as she worked. It dripped on her subject. It dripped on her jacket. It dripped on my notepad. It dripped on her lens. She said, "Stop dripping on me!"

By four thirty, there was time for only one more shot. She made her way toward the formerly cobalt panel. The best vantage point, she determined, was tricky to get to: up a flight of stairs, down a ten-foot ladder, over a grate, and onto a four-foot pipe. From there, she'd traverse forty feet out, using smaller gas pipes as railings, and then follow the pipe where it bent up at 30 degrees. Following behind Csük, I reached down from the top of the ladder to pass her the tripod. The big pipe reminded me of the Alaska pipeline. It was the same size, about as snowy, but it was thirty feet up above concrete and steel instead of four feet above tundra. From out on the big pipe, the panel appeared gray and mottled.

"It's subtle," Csük said. It was the first time I couldn't tell what, exactly, she was aiming at. "See how beautiful it is?"

She took a few photos. The tripod kept slipping.

"I like it. This might be better than what I shot before."

She climbed a bit higher, took another photo, and struggled to brace the tripod on the slippery snowy cylinder. Then she said it was too dark, so we headed down.

Csük led the way back through the scraggly littered courtyard to the spot at the chain-link fence where we had climbed in. On the way there, she caught her foot on some rusty thing and tripped. She caught herself easily. It was the only time I saw her trip.

Though it was dusk, it was not yet dark enough for comfort. Getting out turned out to be tougher than getting in.

I climbed over the fence first. Once over, Csük passed me the tripod. Then she saw lights approaching from the east.

"Walk!" she said.

"Where?"

"Just walk!" she said. "Go! Anywhere!" Then she retreated into the steelyard.

I thought about climbing back over the fence, but wasn't sure I could do it fast enough. So I hurried away from the lights. I stuck along the fence, looking for spots to hide. There were none. The saplings were all too thin and leafless for cover. I looked back: the trio of lights was getting closer. I considered lying down and staying still. Then I saw a buttress of steel girders, and hid behind it. I felt like a cartoon character squatting behind a too-small bush. The sound of the approaching train grew louder, the shadows on the ground shifted, and then the train—only two engine cars—was gone.

I jogged back and met Csük at the fence. Wordlessly, she passed me her backpack and then climbed over. She made it fewer than ten feet before she said, "Did you hear that?" I had heard something, but it sounded distant. Then I heard it again. It was the sound of a car on gravel—which meant a security patrol. Though Csük couldn't see anything where the sound had come from, she decided to hustle. She skirted along the fence, and then across the tracks, over the parked train, and onto the levee. I followed.

The sight of the Lehigh River was soothing. I felt free and contentedly weary. My feet were soaked, my pants drenched up to the knees, my hands cold. As I walked the half mile behind Csük, I thought about a hot shower, the only way to extract the rust dust from my hair. Then Csük froze. The train that had gone by was stopped under the bridge, exactly where we wanted to cross. Beside the train, guiding it into position, were two railroad employees. "Shit," Csük said. "You do not wanna mess with those guys."

Alyssha Eve Csük—so experienced in patiently extracting beauty from this unfriendly, forsaken place—decided that the most prudent thing to do was make a beeline for *terra publica*. She slipped down the snowy grass, crossed the tracks, and looked both ways.

Then she ran.

6

THE AMBASSADOR

Dan Dunmire arrived in Kissimmee late and looking something awful. It wasn't quite ten in the morning on the day before Halloween, and as Dunmire stepped out of his Ford Excursion into the small parking lot at Bruno White studios, his face, hidden beneath a Pittsburgh Steelers cap, appeared haggard. His eyelids sagged, and his thin gray beard looked matted. He scuffled more than he walked. Nothing about him suggested Pentagon official. His black slacks hung low, and his khaki button-down shirt ballooned out from a large waist. Exacerbating the ballooning effect was a thin nylon Steelers jacket. Velcro shoes and white socks bearing the Steelers logo completed Dunmire's disheveled look, but Dunmire—sixty years old and proudly quirky—didn't care. He's a Pittsburgh fanatic, always wearing, beneath business attire, at least one article of clothing from the city. His exhaustion and tardiness were typical, too, but their cause was not. He'd been driving for fifteen hours, through Hurricane Sandy, which had grounded the flight from Pittsburgh that he'd planned to take. Through horizontal rain and eighty-mile-per-hour winds—through seven states, three of which had declared states of emergency, through a natural disaster that stopped Romney, Biden, Obama, and Clinton from campaigning—he'd driven, with only a short stop outside of Washington,

DC. To Florida's clear skies this federal official drove because he was determined not to miss the day's video shoot. The only thing that Dan Dunmire, the nation's highest-ranked rust official, likes more than Pittsburgh is *Star Trek*, and actor LeVar Burton, a.k.a. Lieutenant Commander Geordi LaForge (the officer with the *thing* over his eyes), was in the dressing room getting ready to talk about rust.

Just inside the front door, Dunmire grabbed a bottle of water from a buffet table and proceeded through the director's room into the studio proper. He took a handful of pills, a swig of water, and gulped. "I feel horrible," he said hoarsely. "Lousy. 5-hour Energy, yeah! I had three 5-hour Energies in less than fifteen hours. Yes!" Burton emerged to greet him. Tall and confident, he was wearing a navy hoodie and fashionable jeans, sporting silver-tipped black leather shoes, a scarf, and a mustache two degrees slicker than that of any engineer. He said, "Dan the man." They shook hands and hugged, and Dunmire's weariness evaporated.

Dunmire's official title is director of the Department of Defense's Office of Corrosion Policy and Oversight, but he has called himself the corrosion czar. It's easiest to think of him as the rust ambassador. In his dealings with industry, academia, and the military, implementing hundreds of rust-prevention measures, that's a fair representation of the role he fills. In enlightening and educating the public about corrosion—a large and unique ambition of his—the rust ambassador has ceded much of his position to LeVar Burton. The actor, with two decades of experience hosting the children's TV show *Reading Rainbow*, is the rust ambassador's rust ambassador. He is the Pentagon's public face for rust.

Since 2009, Burton has hosted four thirty- to forty-five-minute Pentagon-funded corrosion videos for a series called *Corrosion Comprehension*. The first, *Combatting the Pervasive Menace*, defines the challenge facing the Department of Defense and the country. Demeaning rust as "a defiant and dangerous enemy," a "silent, pervasive, and unrelenting scourge," and a "real and present danger," Burton sounds the alarm: "It's what we can't see that's even more troubling." What Burton, Dunmire, and the rest of the DOD can plainly see and be troubled by is rusty plumbing in aircraft hangars at Fort Drum, rusty bridges at Fort Knox, rusty heat pipes at the Redstone Arsenal, rusty roofs at the Kilauea Military Camp, a rusty water treatment system at Fort Huachuca, a rusty water tank at

Fort Lewis, rusty pumps at Fort Polk, rusty diesel tanks at Fort Bragg, rusty masonry straps at Fort Stewart, rusty munitions storage facilities in Okinawa, rusty fire hydrants, rusty air-conditioning coils, rusty jeeps, rusty tanks, rusting jets, helicopters, missiles, cruisers, and aircraft carriers.

In the second and third videos, Burton focuses on corrosion prevention via polymers and ceramics, deploying long strings of factual but numbing passive sentences (". . . in this case, the continuous matrix is an organic polymer such as epoxide or polyurea . . .") that bring to mind an audio-book of a medical journal or *Star Trek* gibberish. The fourth video brings the subject back to earth by examining how the military operates in various corrosive environments, much like a top-ten list. Considering the videos outreach and equally valuable to the public and the military, Dunmire has posted them on the Office of Corrosion Policy and Oversight's website. Always involved in the script writing, Dunmire has allowed that they have their elements of kitsch. Burton's introductions ("Hi, I'm LeVar Burton, and I'm going to take you on a journey . . .") are no less cheesy than his closings (". . . and remember, rust never sleeps"). *Star Trek* and *Reading Rainbow* insinuations lighten things up, but sterile techno/Enya music and Burton's soothing voice, sinusoidally intonating like that of an NPR reporter, have a soporific effect. The exposition-to-action ratio runs pretty high. The videos were Dunmire's attempt to bring to life the unglamorous subject of corrosion. Whether or not they have succeeded is debatable. Regardless, Dunmire drove to the studio to oversee the production of a fifth corrosion video, *Policies, Processes, and Projects*, which he calls "LeVar 5." The subject of LeVar 5 is the Office of Corrosion Policy and Oversight and its "aggressive" plan to fight corrosion. In other words, Dunmire came to Kissimmee to oversee the production of a movie about himself.

While a dozen staff were readying for the shoot, and Burton was changing, Dunmire greeted the executive producer, a woman named Lorri Nicholson. Around her, an audio guy, a lighting guy, a jib guy, a makeup artist, a teleprompter, a photographer, a caterer, two cameramen, and two production assistants shuffled about. Wires snaked every which way on the floor. Dunmire rocked back and forth on his feet as he caught up with Nicholson, whose composure only highlighted Dunmire's swaying. Dun-mire sways when he's excited or exhausted, which is most waking hours, and it makes him look nothing so much as kooky. He tilts his head down, locks eyes, lets his arms hang at his sides, and teeters. It does not put new-comers at ease. People who know Dunmire, though, are used to it. Unfazed were Stacey Cook, the cheery producer of the shoot, and Shane Lord, the mellow creative director, both of whom greeted Dunmire warmly. Lord, in ripped jeans, had pulled his long hair back in a ponytail and removed his shoes, so that he skirted about the studio silently in argyle socks. Cook, for the occasion, had put on a short-sleeved shirt bearing the unofficial and oversized logo of Dunmire's office. It said Corrosion Prevention and Control.

Dunmire shuffled past the teleprompter, the boom mike, two Kino lights, and three video cameras—one on a jib crane—and stopped at the edge of the green screen. There he settled into a folding chair, donned a pair of headphones, picked up a clipboard, and began examining the twenty-five-page script. Dunmire had to make sure that everything LeVar Burton said would pass DOD muster. The DOD, decidedly not mov-ing at warp speed, had not yet approved the script for LeVar 5, and the script, written by a man in Los Angeles, was riddled with minor flaws. Dunmire—in the Pentagon since Reagan's first term—was there to clean up the script as he suspected the DOD would. In his chair, he was still the rust ambassador, but he was also an executive producer.

At ten o'clock, Cook yelled, "Cell phones off!" She closed the doors, turned on the lights—and the studio began to heat up. The talent entered. Burton now sported a black suit over an unbuttoned striped shirt with no tie. As he walked into the studio, he was talking with Lord about zombie movies, and the first thing that everyone heard him say was, "In the case of a zombie apocalypse, stay the fuck out of San Francisco." To ears that grew up on *Reading Rainbow*, it was a shock. He walked onto the green screen and took his position where Lord directed him. Dunmire, with the script in his left hand, reminded Burton, "We're trying to get the word out." Bur-ton said, "Stem, baby, stem," by which he meant STEM, for science, tech-nology, engineering, and math. Then he asked Lord, "Are you sure zombies don't eat soup?"

While one of the PAs covered the silver tips of Burton's shoes with black gaffer's tape, the actor rubbed lint off of his suit and practiced in-toning the lines of the script from the teleprompter. Dunmire, ten feet away, sat with his legs crossed, a cup of coffee in his hand, rapt. From his front-row seat, his eyes were glued to Burton in pure idolatry. The baby boomer resembled any of the boys up the road at Disney World.

With the cameras rolling, Burton began from the top of the script: "The United States is a nation that earns a proud place on the world stage," he said mellifluously. "Teeming with activity are its bustling cities, its busy highways, a proud military, and magnificent landmarks. And yet, a subtle and silent enemy threatens it all." Here his voice got suggestive. "For every gleaming bridge, there's another near collapse. Buildings decay, pipelines explode, roadbeds crumble. The enemy is rust." Curiosity oozed from his voice. He continued, pacing across the floor. "It's a destructive force that costs us over half a trillion dollars a year. With that much at stake, a new initiative is under way, with Washington, DC, as general headquarters of an all-out war on corrosion . . . Corrosion, you see, whether on iron or any other material, is something that never stops. We can fix it when it happens, and we can try to prevent it, but all we can ever really do is slow it down. That's why we call it"—here his voice lowered—"the pervasive menace."

Lord looked to Dunmire and asked, "Okay?"

Dunmire piped up, "Wait, we gotta change that."

Burton said, "What?"

Dunmire said, "About half a trillion, not over half a trillion. If we go over, it's not right."

Burton did a second take, and kept going after the bit about the pervasive menace: "I'm LeVar Burton. Join me now as we meet some of the people who have come up with an innovative and successful plan to fight this insidious peril. Watch as they lay out the battle strategy to fight corrosion as no one has ever fought it before." He finished the take and then, in an aside to Dunmire, said, "Subtle. Nice. Boom!"

Just to be safe, Burton did a third take. Partway through, he stopped abruptly after saying, "For the safeguarding the national security of the United States." The word *of* was missing. To himself he said, "How did I get through that twice?" Dunmire leaned back and grimaced, fixed the script, and had the teleprompter make the edit. While Burton practiced the next section, makeup came out and patted sweat from his face.

Burton went through the next section. "Congress chose the Department of Defense to lead our war on corrosion," he said, "at least in part because the army, navy, marine corps, and air force have so much that's vulnerable to corrosion." Dunmire interrupted. "Stop, no, no, no," he said. "The marine corps is not a department. They're a service, so they come last." Burton raised both eyebrows, and after the edit was made for the teleprompter, he resumed. The cameras were still rolling. He said, "army, navy, air force, and marine corps," in the proper order, and proceeded.

In 1998, at the request of the Department of Transportation, the National Association of Corrosion Engineers began estimating the cost of corrosion. By the summer of 2001, NACE figured that the military cost alone amounted to $20 billion. Before NACE published its study, it sent Cliff Johnson, the organization's public affairs guy, to Washington to see about taking action. The way NACE saw it, just applying what we already knew could save six of those billions. At the US Senate Armed Services Committee, Johnson was directed to the staffers on the Subcommittee on Readiness and Management. There were two staffers. One covered military construction. The other, who covered everything else, had just arrived on Capitol Hill. Her name was Maren Leed, and though she was only thirty years old, she was a former fellow at RAND Corporation, the defense-oriented think-tank, with a doctorate in quantitative policy analysis. As it happened, she'd also worked as an analyst in the Office of the Secretary of Defense (OSD) and was hunting for a signature issue for the new ranking member on the committee. That was Senator Daniel Akaka of Hawaii— a state very much victimized by military corrosion.

Looking into military corrosion, Leed found "a huge cost-sink for the department that nobody talks about or cares about." She became convinced that incentives at the department were set up wrong, that institutional biases made fixing the problem nearly impossible, and that the issue wasn't getting appropriate attention. Inquiring with the DOD's Office of Acquisition, Technology, and Logistics (AT&L) about the matter, she got no reply. The way she saw it, in response to corrosion, the DOD was doing bubkes. When corrosion matters in the Pentagon were handled, they were distributed to people in four different offices, in logistics, research, engineering, and infrastructure, but to nobody in the OSD. The branches of the military were even more scattered. Navy officers in one ocean didn't know what navy officers in another were doing. Sailors working on similar problems didn't know each other and certainly didn't know about similar problems in other branches. For all the navy knew, the army might have had aircraft carriers.

At a committee-wide meeting, when colleagues around the table pitched initiatives about acquisition reform or Afghanistan, Leed proposed corrosion. Akaka liked it and got behind it. In March 2002, Senator Akaka officially proposed creating a separate corrosion officer in the Office of the Secretary of Defense. He proposed that this corrosion officer be a political appointee, confirmed by the Senate, reporting directly to the undersecretary. The officer would need to compile databases of military projects and operations to track his or her progress. During debate in the Senate on the matter, while Akaka got up to speak, Leed's colleagues poked fun at her. In behavior that was, at least back then, unbecoming of the chamber, they circled their hands around their heads, forming crowns. They called Leed the Queen of Corrosion. It was captured on C-SPAN.

Word spread. Appropriations committees got on board, as did Senators Thad Cochran (Mississippi) and Daniel Inouye (Hawaii), and Representatives Betty Sutton (Ohio), Bill Shuster (Pennsylvania), and Rob Portman (Ohio). They represented the rust belt well. California perked up too.

Inside the Pentagon, the proposal wasn't received with such fanfare. Michael Wynne, the recently confirmed deputy undersecretary of defense for AT&L, thought that the reporting stipulations of the proposed law were overly burdensome. He and other DOD senior leaders figured the corrosion proposal was entirely the dream of the young committee staffer. They accused Leed of the cardinal sin of Senate staffing: getting in front of her rep. Leed recalls that Wynne was "really pissed" to the point of hostility. He arranged to meet Akaka. When Akaka—the opposite in temperament—couldn't make it, Wynne met with Leed instead. She bore the brunt of his fury. "He said, 'I'm running two-hundred-million-dollar programs here, and I don't have time for this crap,'" she recalled. She wondered, "Who doesn't have time to save millions of dollars?" Between Leed and Wynne, Dunmire—who had been busy writing briefing reports for testifying officials—brokered a settlement.

Dunmire spent that summer and fall navigating the terrain between the Armed Services Committee and the undersecretary, reviewing draft statute language that was feasible to both. Way down on page 200 of Congress's annual defense authorization bill—a hefty three-hundred-pager loaded with post-9/11 amendments—that language appeared. It stipulated the designation of a senior official responsible for the prevention and mitigation of corrosion of military equipment and infrastructure.

The House and Senate agreed on a conference report on the bill in November, and sent it to President Bush's desk two days before Thanksgiving. Bush signed it a week later. The Bob Stump National Defense Authorization Act didn't appropriate an office or any funding for this new corrosion executive, and it further annoyed Wynne by giving him ninety days to pick someone.

Dunmire certainly didn't want the position. Though he figured it could lead to phenomenal changes, he pitied the chosen occupier of the new chair. He felt sorry for the sonofabitch who got the job, thinking the task close to impossible. He could imagine so many people not suited for it, and with Ric Sylvester, a longtime senior executive in the AT&L office, used to laugh his ass off thinking about how hard the job would be. He never imagined he'd be a part of it.

Figuring he could perform the duties of the undersecretary *and* the corrosion executive, Wynne picked himself for the job. To manage the program, he decided to name a task-force director. At the Pentagon, he called in his staff. He dismissed everyone but Dunmire.

Dunmire struggled through his first year as the sole staff member of the new corrosion "office," which kept moving around northern Virginia: from the Pentagon, to Alexandria, to the Pentagon, to Crystal City, to the Pentagon, and back to Crystal City. With no budget, not even a phone, all he could do was develop a strategic plan, a battle strategy. Dunmire was not and is not an engineer. At Kent State University, he studied communications, and at the University of Alabama, he got a master's in public administration. (He's currently earning a doctorate in public administration from Virginia Tech.) So he turned to trained engineers for help. His eventual chief technical engineer, Dick Kinzie, had thrice audited corrosion costs in the air force, and figured the cost of military corrosion was around $9 billion. In 2003 Wynne directed a few hundred thousand dollars toward Dunmire's endeavor, and Dunmire used it to hire part-time help. Finally, in 2005, Dunmire got a budget, albeit not from Congress. Mr. Wynne allocated $27 million to Dunmire, and he used part of it to detail the actual costs of corrosion. He funded studies that broke up the research into chunks. Dunmire recalls that Robert Mason, the assistant deputy undersecretary of defense for maintenance policy, programs, and resources, didn't think such an assessment was possible, or that the resulting numbers would be meaningful, but it was, and they were, with a methodical process. After Mason died, a DOD award in maintenance excellence was named after him; Dunmire often looks at the sky and says, "Hey Bob, I did it." (Of course, the military considers corrosion in the broadest possible sense, as the degradation of any materials—not just metal—on account of ultraviolet light, mold, mildew, or decay.) LMI, the contractor that did the analysis, ran the numbers from the top down and from the bottom up, checking that they converged.

In 2006 Dunmire's office declared that corrosion was responsible for $2 billion worth of annual damage to the 446,000 ground vehicles belonging to the army, and $2.4 billion in annual damage to the 256 ships belonging to the navy. The next year, Dunmire's office declared that corrosion was responsible for two-thirds of a billion dollars annually in damage to ground vehicles belonging to the marine corps; $1.6 billion annually in damage to 4,000-odd aircraft and missiles belonging to the army; and $1.8 billion annually in damage to facilities and infrastructure belonging to the Department of Defense. The year after that, Dunmire's office declared that corrosion was responsible for $2.6 billion in annual damage to the 2,500 aircraft belonging to the navy and the 1,200 aircraft belonging to the marine corps, and a third of a billion dollars in yearly damages to aircraft and ships belonging to the coast guard. Finally, his office declared that corrosion incurred $3.6 billion in damage every year to the aircraft and missiles belonging to the air force. The total was $15 billion: almost exactly in the middle of the two previous estimates.

Engineers in Dunmire's office cite competing incentives as a cause of much of their rust troubles. Program managers in charge of new weapons systems get graded on performance, schedule, and cost. If one oversees the building of a $500 million missile that can fly to Mars by 2015, he's evaluated on the millions spent, the missile, and the date of completion. If the missile rusts to hell in 2016, that's not his problem. Captains or colonels, meanwhile, want stars on their shoulders, and get them by staying on schedule and on budget. To save money, they use cheap fasteners. They use cheap paints, scoffing at new expensive blends that come with vague, optimistic guarantees. "By the time the maintenance bill comes through," they figure, "I'll be gone." It's not hard to race rust and win. Officers get their stars, and assets get treated like orphans.

Congress tried to fix that, too. In 2009 it established corrosion executives in each of the military departments.

After LMI repeated the round of studies, in 2011, Dunmire's office pegged the direct costs of corrosion to the military at $21 billion. Rust, it said, accounted for a fifth to a quarter of the military's maintenance costs. On 162 types of army aircraft, 102 types of navy aircraft, 56 types of air force aircraft, and 31 types of marine corps aircraft, that was a lot of maintenance. They ranked their assets. Corrosionwise, C-5s and C-130s were the worst planes. C-21s were the best. UH-1H helicopters were the worst. UH-60Ls were the best. On marine corps vehicles, corrosion was wrecking alternators, starters, hydraulic lines, underbodies. (About the only thing it wasn't doing was causing parked vehicles to catch fire.) Most of the air force's fleet was aging; bombers were thirty-five years old, tankers were forty-five years old, and as a whole, the average age of the fleet was twenty-five years.

Corrosion interfered with the availability of military assets, too. The Corrosion Prevention Office determined that because of corrosion, each aircraft belonging to the air force was "not mission capable" sixteen days a year. Each aircraft belonging to the army was out seventeen days a year. Each aircraft belonging to the navy and the marine corps was out twenty-seven days a year. Even in Tucson, aircraft rusted (thanks to morning dew). Then there were the freak events: rusted electrical contacts on F-16s that inadvertently shut fuel valves and led to at least one crash; rusted landing gear that mired F-18s landing on aircraft carriers; a rusty bolt that was blamed in the failure of the rotor system in a Huey that crashed and killed an army captain; a rusted electrical box that was blamed in the electrocution death of a sailor. As a result, the air force computed the cost per aircraft, and the cost per pound of each aircraft. On ships—all ships—the problem was tanks: ballast tanks, fuel tanks. Emptying, cleaning, inspecting, preparing, and painting them was frightfully laborious. Plenty of sailors have gone brain dead and/or deaf chipping and painting, chipping and painting, endlessly, across oceans. But their efforts have proven no match for what Steve Spadafora, the navy's sharp, goateed corrosion executive, referred to as "the giant electrochemical cell we call an aircraft carrier." Dunmire regularly calls the problem "humongous," and no admiral would disagree. Studying the cost of corrosion was just his warning shot.

⌒⌒

In the studio, Dunmire asked Burton to take two. "It popped or something," he said. "Something went wrong. Could you do it again?" Burton, reading from page 3, said, "All that iron and steel must be constantly protected from rust."

A few takes later, just as Burton was about to start, the audio guy raised his hand. A truck had pulled up outside the studio and left its engine grumbling. Everybody could hear it. Lord said it was the FedEx truck, and waited. Burton took a seat. One of the PAs ran outside and came back reporting that it was the water guy. Cook said, "Tell him to hurry up." Lord, in the meantime, decided to keep rolling.

Burton soon announced, "Let's face it, rust happens everywhere, and it affects everyone. Most of us don't even pay attention to it until something goes wrong. And when it does go wrong, the consequences can be disastrous. Here's what happens when corrosion results in a worst-case scenario. In 1967 the Silver Bridge across the Ohio River between Ohio and West Virginia collapsed. Forty-six people were killed. A single corrosion-induced crack only a tenth of an inch deep was the cause. The Defense Department may be leading the fight against corrosion, but the crusade extends to the entire country."

"Okay, great," Lord said. "Let's do that again."

Dunmire said, "You added the word *even*. Nice add, LeVar."

Burton did another take. "Congress chose the Department of Defense to lead our war on corruption—" He stopped and walked offstage. He meant *corrosion*, not corruption—though either was a formidable enemy. Lord said, "I hate when that happens."

As Burton resumed, the truck rumbled to life and pulled out of the driveway, beeping as it backed up. Burton stopped and looked down. Dunmire removed his headphones. The audio guy raised a hand and held the other to his right ear. He held the position for a few minutes. Everyone stood by. Then a plane flew overhead, and the standing by continued. Eventually Burton resumed.

"Modern history is littered with corrosion disasters," he said. Lord said, "Great, that's it." Yet the gravitas of the disasters failed to capture anyone in the room. It wasn't great at all. Burton's tale—as convincing as any story about rust could be—was just another sales pitch, like any of the infomercials produced in the studio.

If the script sounded like a blend of evening news, *Entertainment Tonight*, and *National Geographic*, it's because Darryl Rehr, its author, worked at all three. He's covered Rodney King and O. J. Simpson, tornadoes and train crashes, gas explosions and gamma ray bursts.

Summoning more concern, Burton said, "The annual cost of corrosion to the United States amounts to three-point-one percent of our gross domestic product. In the year 2011, that came to something like four hundred and eighty billion dollars a year—more than fifteen hundred dollars for every man, woman, and child in the country, over six thousand dollars for a family of four! Ouch." He put a lot of emphasis on the word *ouch*. Dunmire liked it, and echoed, "Ouch."

As someone blotted the sweat from Burton's face, the actor told Dunmire that the script was scintillating. Dunmire laughed, and then Burton laughed as he repeated the word *scintillating*. At the top of page 8, Burton began explaining the 2002 law that created Dunmire's office. "Congress . . . added a section to the United States Code . . . and the new section . . . starts with this simple phrase: 'There is an office of corrosion policy and oversight.'" Burton gave it as much dramatic emphasis as one could grant a legislative act and then added, in an aside, "words that reverberated around the world." He looked at Dunmire, and they both laughed. He read it again, and after Lord said, "Cut," Burton looked at Dunmire and said, "simple, yet *powerful*!" Burton was poking fun at Dunmire and the script, and Dunmire didn't care.

From the same page, Burton practiced the next section, reading out loud as he explained the new federal code: "The corrosion program would be designed to take what was called an overarching and integrated approach to a problem that everyone was used to dealing with on their own. From now on, things would be different." He said this last part with emphasis and then added, "From now on, there's a new muthafuckin' sheriff in town."

Then he did the real take, omitting the "muthafuckin' sheriff."

Afterward, Burton asked, "Is this a two-act play?"

After a century of unmitigated corrosion, Lady Liberty had developed cracks, scabs, stains, holes, and rust boogers—here investigated by architectural historian Isabel Hill on March 28, 1985. *(Photo by Jet Lowe, courtesy Library of Congress, Prints & Photographs Division, Historic American Engineering Record—HAER NY, 31-NEYO, 89–180)*

A famous Yosemite climber once said that he wasn't sure if Ed Drummond, here in Scotland in 1970, was a poet who climbed or a climber who wrote poetry. His 1980 stunt climb on the Statue of Liberty sparked the most dramatic corrosion battle in American history. *(Photo courtesy Ed Drummond)*

In renovating the Statue of Liberty, the National Park Service sought the advice of Robert Baboian, who ran the corrosion lab at Texas Instruments' office in Attleboro, Massachusetts. At the end of 1983, he began measuring the thickness of the patina on Lady Liberty's copper. Based on his results, he figured her skin would survive a thousand years. *(Photo courtesy Robert Baboian)*

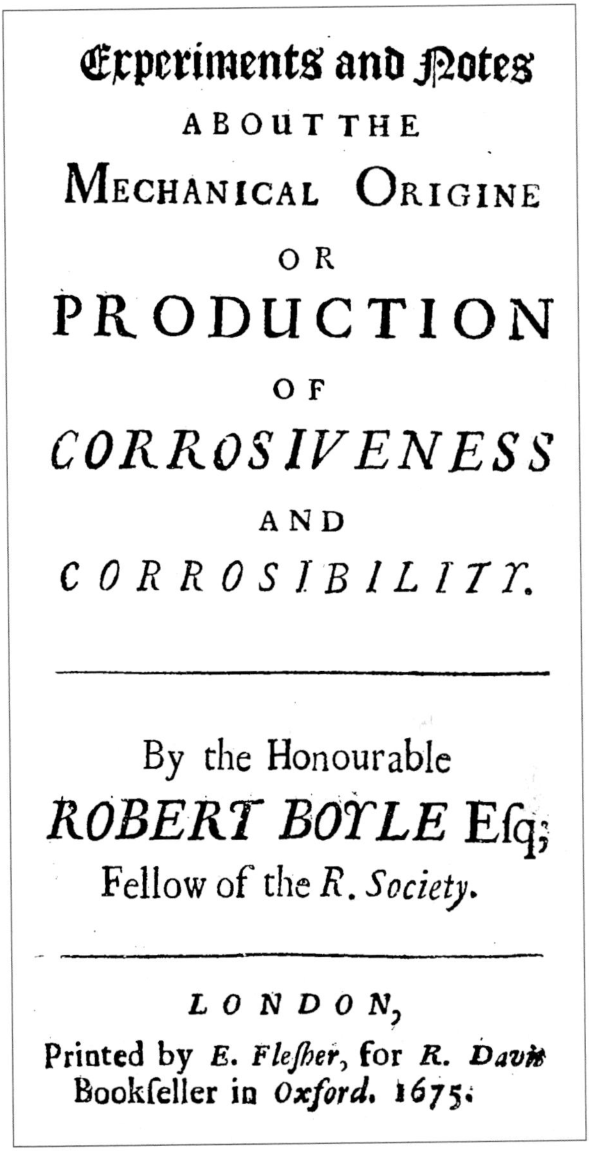

𝔈𝔵𝔭𝔢𝔯𝔦𝔪𝔢𝔫𝔱𝔰 𝔞𝔫𝔡 𝔑𝔬𝔱𝔢𝔰

ABOUT THE

MECHANICAL ORIGINE

OR

PRODUCTION

OF

CORROSIVENESS

AND

CORROSIBILITY.

By the Honourable

ROBERT BOYLE Efq;

Fellow of the *R. Society.*

L O N D O N,
Printed by *E. Flesher,* for *R. Davis*
Bookseller in *Oxford.* 1675.

During the reign of King Charles II, Robert Boyle, the wealthy English "father of chemistry," took up the first proper investigation of rust. His experiments with saltwater, lemon juice, vinegar, urine, and various acids showed that all metals, even gold, were vulnerable. *(Image courtesy Cambridge University Library and Early English Books Online/ProQuest)*

Rust photographer extraordinaire Alyssha Eve Csük headed into Bethlehem Steel for the 377th time in November of 2012. Over the last decade, she's taken nearly 30,000 stunning shots of rust. *(Photo courtesy of the author)*

Dan Dunmire, the Pentagon's corrosion ambassador and nation's highest-ranked rust official, has chosen LeVar Burton to wage a public campaign against rust, which costs the Department of Defense $15 billion annually. Here, in a Florida production studio in October of 2012, Dunmire suggests changes to the script of the fifth of seven corrosion videos Burton has hosted. *(Photo by Diana Zalucky, courtesy U.S. Department of Defense)*

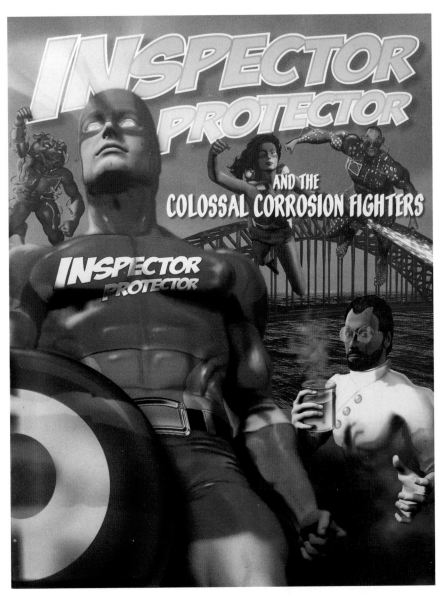

In 2004, the National Association of Corrosion Engineers published an "action-packed comic book adventure" to introduce eight- to fifteen-year-olds to the exciting world of rust. At the group's 2013 conference, an actor dressed as the book's superhero, Inspector Protector, posed for photos with NACE members and officials when not doing flips. *(Image by Marion Integrated Marketing, courtesy of NACE-International)*

In March of 2013, Bhaskar Neogi began a monthlong smart pig inspection of the Trans-Alaska Pipeline, one of the heaviest metal things in the Western hemisphere—and the least accessible. On the day he launched the pig, it was a balmy ten below in Deadhorse, Alaska. *(Photo courtesy of the author)*

Howard EnDean, the father of modern rust-detecting pipeline pigs, had oil in his blood. On one day in 1956, EnDean filed patents for four varieties of smart pigs; one was at least fifteen years away from technical feasibility. *(Photo by James Streiner, Gulf Oil Public Relations Department, courtesy Howard EnDean, Jr.)*

John Carmona, proprietor of The Rust Store in Madison, Wisconsin, collects rusty stuff for product research. *(Photo by Colleen Carmona, courtesy The Rust Store)*

Some of the world's first stainless steel—before it was called as much—was used to make the hull of *Germania*, which was renamed *Half Moon*. Here, off New York City in 1921, she was nearly pristine. A century later, the remains of the storied German schooner now rest—very much stained—in shallow water off Miami, Florida. *(Photographer unknown, courtesy Marjory Stoneman Douglas Biscayne Nature Center)*

Lord said, "It's a five-act play."

Cook corrected him: "Six acts."

They'd just finished act one. More than two-thirds of the script remained.

Burton said, "Who writes six acts? Ugh."

Since Dunmire began officially fighting rust, he's had one agenda for one ultimate purpose: he wants the military to evolve from its current find-and-fix reaction to a form of proactive management, to the benefit of the American warrior. That's his prime directive. (It's the same agenda that Francis LaQue, a founding father of corrosion studies, pushed a generation ago, when he worked as Richard Nixon's deputy assistant secretary of commerce for product standards.) Unlike many engineers, Dunmire recognized that a new approach to corrosion represented a cultural rather than a technical shift. He saw combatting corrosion as a societal concern. The first time I met Dunmire, at a navy corrosion conference in Norfolk, Virginia, called Mega Rust, he told me about wanting to change American military and civil culture. This was in 2009. "Manufacturers *want* stuff to break," he said. "They design stuff to break. At the navy, we don't want stuff to break. So we're looking at corrosion from two different lenses."

He went on: "This is the USA. We're a capitalist society. We cannot afford for the DOD to be a profit center in the business of corrosion. We have to break this cycle." He said, "I'll pay for it, but I don't wanna pay for it over and over and over again. Gimme the good stuff the first time. Make sure it works, and make sure it lasts." He told me about developing a national anticorrosion culture, focusing on prevention, and training more engineers and scientists. Hence his love affair with STEM programs. He wants to get kids turned on to corrosion, or at least engineering. "It's insidious and pervasive, but not inevitable. It can be predicted, prevented, detected, treated," he said. Saving 30 percent wasn't good enough. "We gotta go from find and fix to predict and manage. Otherwise it's twenty billion dollars, and that makes me upset."

His feelings seem to have rubbed off on a few admirals. Vice Admiral Kevin McCoy, the revered commander of Naval Sea Systems Command and also a mechanical engineer who trained at MIT, said in his keynote speech at Mega Rust that he thought corrosion would destroy the US Navy. Congress had recently approved expanding the fleet to 313 ships, but McCoy saw that aim as unfeasible. "It's great to buy new ships," he said, "but we can't buy our way to three thirteen. We gotta keep three-quarters of that with our old ships." His great fear was that corrosion would bring the US Navy's fleet down to 200 ships, and his assessment of where the navy stood was not good: "We are completely blind right now." Rear Admiral Thomas Moore, the deputy director of fleet readiness, agreed with him. Rear Admiral James McManamon, the deputy commander for surface warfare, said, "I wanna be able to decide when to decommission a ship, not have the ship decide." Rear Admiral John Orzalli, the commander of US Fleet Forces, said, "If we just keep doing more today of what we've been doing, we just don't have the funds. Just doing it the way we've always done it won't work." Asked how sailors could be taught about corrosion, he said, "We need to be developing a questioning attitude." That was provocative stuff for a military that prides itself on a rigid, top-down command structure, with everyone bent at one knee. Dunmire's campaign was taking hold.

Dunmire, admittedly headstrong and opinionated, admires bold leaders with lofty ambitions, regardless of their political foundations—men such as Admiral Hyman Rickover, who developed nuclear-powered submarines, and Wernher von Braun, the Nazi turned NASA aerospace engineer who fathered rocket science. He admires underdogs, like Confederate commander John Singleton Mosby and Union brigadier general Joshua Lawrence Chamberlain. Generals George Patton and Erwin Rommel both make his list. In his war against rust, Dunmire has pushed up against the boundaries of acceptable behavior. All the while, he's remained patient, aware that "fighting the second law of thermodynamics," as he regularly puts it, requires resiliency. Perhaps his quirks have helped him remain steadfast where others would have given up. He says things like, "It's fun taking on God"—then clarifies that he meant it as a joke. Another time, in the same vein, he once said, "In the *new* New Testament, we're gonna talk about corrosion. To God, I'll say: you win, but we need a little comic relief."

Or he'll say, "We will continue to fight the good fight." The good fight, as he sees it, is getting corrosion into the minds of everyone in the military: from high-level decision makers, including the secretary of defense, to first-class servicemen and the officers and bureaucrats in between. But it's a tough campaign. Dunmire knows that adopting his agenda won't be easy, but he has said that it's an easier fight than gun control or terrorism. He tells people not to look at fighting corrosion as a negative thing but as an opportunity to make materials last longer. "We're planting seeds," he said. "In my life, I don't know if it'll succeed." He always insists that he's not done, that his program is growing as incrementally as rust.

Actually, it's dealing with human beings that require even more patience than dealing with aircraft and ships and bases. Physics he calls black and white. People he calls "quasi scientific at best." They have more momentum than a carrier and require as much space for making turns. Figuring that psychological battles must be fought in person, Dunmire has traveled a lot in order to, as he says, connect the dots. Since 2005, his work has taken him to Guam, Japan, South Korea, the Philippines, Australia, Germany, England, Italy, Alaska, Hawaii, Texas, Nevada, and Florida. At Hilton hotels, he was a diamond member. At the hotel near Kissimmee, both bartenders knew Dunmire's name and poured him scotch on the rocks without asking what he wanted. Over the last five years, he's spent at least one of every four weeks away from the five-sided windowless vault known as the Pentagon. He regularly visits Carderock, Patuxent, Norfolk, Aberdeen, Quantico, Fort Belvoir, and Andrews Air Force Base.

With officials and officers at such places, he's not afraid to tell them they screwed up. He tries to be tactful but firm. "Make your point and get out," he says. "But if someone disagrees, you know what you say? You say, 'Yes, sir.'" The rest of the time, he's on the phone with people at such places. "You can't manage this program from a desk," he's said. Thrice annually, his office holds a corrosion conference, naming them like Super Bowls. For Corrosion XXXI, a month after LeVar 5, he hoped to catch up with dozens of players in the rust world. The government did not like the sound of that.

In October 2010 the General Services Administration—the government's secretarial arm—sent more than three hundred people to its four-day Western Regions conference in Las Vegas. Attendees stayed at a four-star resort, ate $44 breakfasts and $95 dinners, and enjoyed the services of a clown and a mind reader. The total came to 8 million government dimes. When the largesse was discovered and publicized in April 2012, the department chief and deputy were forced out, and GSA's administrator resigned. Then the Office of Management and Budget tightened the rules governing conferences and symposia. Half a year before Corrosion XXXI, any meeting between more than two people was henceforth defined as a conference, and conferences became no-no's.

Determined to talk about rust woes face-to-face, Dunmire renamed his corrosion conference. It became Corrosion *Forum* XXXI. Two dozen people came, fewer than expected, prompting Dunmire to call it a forum-minus. There Dunmire announced that in 2014 he would hold a praxis. Since there was no registration and no fee, DOD wouldn't consider it a conference. In any case, whatever it was, it was as bare bones as could be, in a bland room on the third floor of a building in a Beltway office park. There were no clowns or mind readers or catered meals, only chairs and endless PowerPoint presentations and cafeteria trays.

At the gathering defined by creative hairsplitting, Dunmire was irked that all of government was punished for the mistakes of a few. To those who'd made it, he said, while rocking on his feet, "You gotta do this face-to-face, rather than videoconferences, counter to federal policy. I'm guarding jealously the ability to meet. We're meeting with general counsel and WHS [Washington Headquarters Services] to make sure we can continue to meet. It's not that we wanna get out of our office, it's that we can't do our job from the office. So thank you."

~~~~~~~

The second act was to begin, with Burton grasping a chunk of iron as he roughly explained the principle of entropy. Holding the twenty-five-pound rusty prop, Lord said, "Where is this rock gonna come from? Out of his ass?" Burton laughed loudly. Lord said, "Can it sit on a stool? Not that kind of stool!" Then he told the story of Shaquille "Shaq" O'Neal, who, in the studio the day before, had taken a photo of his egesta and showed it to everyone.

Poop references were fresh. Dunmire, in fact, had just told me that a few weeks ago, at an accelerated corrosion facility in Nevada, he dropped his cell phone in a toilet. Since it was government property, he returned it to the Pentagon. A guy there put it in a bag labeled "Dunmire. Toilet." Dunmire is not above poop jokes. Referring to the Facilities and

Infrastructure Corrosion Evaluation Study (FICES), he calls it feces. "Based on the findings of the *feces* study," he told attendees at Corrosion Forum XXXI, "you'll have a lot to chew on." Similarly, the navy's Shipboard Corrosion Assessment Training course is supposed to be called and pronounced S-CAT, not SCAT. Dunmire takes humor where he can find it. In the early days of the corrosion office, when it was just him and two sidekicks, he called the bunch the three rusketeers. He's since tried a few corrosion jokes and confirms that not one is funny. At NACE's 2012 annual corrosion conference, he told me, "My jokes never go over; I have to say, 'I just told a joke.'"

Just then, another truck showed up, grumbling—and everyone took five. It turned out this truck was full of cases of crème brûlée—a pallet's worth, because Nicholson also runs a food business. So half the crew, including Lord, in his socks, began unloading the truck and stacking the boxes in the shade.

Burton passed on the manual labor. He stepped out of the lights, took a seat in the director's chair, crossed his legs, and rolled his second cigarette of the morning, from a pouch of Bugler Turkish tobacco. When he finished, he went outside to smoke in the shade of the trailer. There I asked him how Dunmire persuaded him to do these corrosion videos. "It didn't take a lot of convincing, once I knew he wasn't a stalker," Burton said. Then he added, "The military-industrial complex is obsessed with *Star Trek*."

In his first years in the Office of the Secretary of Defense, when Frank Carlucci was the boss and Dunmire was just a GS-9, Dunmire attended his first *Star Trek* convention. He went dressed as a civilian, brought his wife, and met James Doohan, who'd played Scotty, the engineer in the original series. In the summer of 2006, a few rungs up the Civil Service pay scale, Dunmire decided it was time for another. To the big one, the fortieth-anniversary *Star Trek* convention, he headed. Fifteen thousand Trekkies poured into Las Vegas and immersed themselves in science fiction for four days. Dunmire's wife stayed home, but Dunmire (dressed as Captain Jean-Luc Picard from *Star Trek: The Next Generation*) brought his fourteen-year-old son (dressed as the same). They stayed at the Stratosphere hotel. The convention was a mile south, at the Las Vegas Hilton.

Twice a day, this Pentagon official with top-secret clearance—whose boss's boss reports to the president—walked down Paradise Road with the emblem of the United Federation of Planets' Starfleet Command on his chest.

At the convention, Dunmire sat stage left, halfway down. Beside him was a guy who'd biked there from California, brought no other clothes, costume or otherwise, and smelled like it. On the second day, while Worf was onstage, a woman (dressed as Dr. Beverly Crusher) sitting just past the stinky biker guy turned toward Dunmire and told him to shut up. Dunmire had been talking for two days straight about you know what. Recalled the woman: "I was, like, 'Dude, you're obsessed with the Kyrosians, they were only in two episodes, and I'm trying to watch Worf!'" Dunmire said to her, "No, not Kyrosian—*corrosion*. I'm telling you what I do!" They began talking about Dunmire's job, his war on rust. The woman—Stacey Cook—revealed that she was a video producer. Dunmire mentioned his intent to make corrosion-related videos. Cook reached into the tricorder that was slung around her shoulder. Beside a medical scanner, she had a stack of business cards. She and Dunmire exchanged cards. Dunmire called within a month.

During the next year, Cook produced a few short video podcasts for Dunmire's office. Dunmire, as the head of the office, was the talent. He dived into the role like it was his first day on a ship. In one video, he pops up on screen, wincing quizzically, as if giving the camera a hard time. He dances, sort of. In another, five versions of Dunmire, all dressed differently, are sitting in a room, watching a sixth Dunmire talk about corrosion on a screen. In another, he picks up a globe while pondering, in the manner of Orson Welles, "How am I going to get this message across?" The scene dissolves in a dream sequence to open-mike night at the Pentagon, where Dunmire is a stand-up comedian. He's got a lit cigar and a cue card, and he's talking about Congress's corrosion mandate. Describing these podcasts, Dunmire said, "We're trying to be humorous!" He added, "It's self-deprecating humor, but it keeps people watching." They're also tax dollars, uniquely spent.

Still, Dunmire wanted to go bigger. He wanted to make a video to succeed the classic but outdated (and boring) 1954 educational film *Corrosion in Action*, which features crude and inaccurate animation of a corrosion cell. Dunmire was unwilling to pay LaQue $50,000 for the rights to it; he could paint part of a plane for that price. The video idea idled until Dunmire and Cook returned to Vegas for the 2007 *Star Trek* Convention. Dunmire went in the original series gold shirt as Captain James T. Kirk; Cook went as Nurse Christine Chapel. Her regular employer, Lorri Nicholson, came along for the ride even though she thought both were big geeks. She was right: Dunmire knows and regularly says the middle name of Captain James T. Kirk. In Vegas, Dunmire heard LeVar Burton say that he was looking for work, and that's when it clicked. Having hosted *Reading Rainbow*, Burton had the educational credibility and prominence that Dunmire was looking for. Nicholson and Cook, who had filmed a TV production for Disney and NASA with LeVar Burton, said, Well, we know LeVar. Two months later, Dunmire sent a letter to Burton via his agent. Four months after that, the night before production began on LeVar 1, Dunmire and Burton finally met in Florida. After a few drinks, Dunmire followed LeVar Burton into the men's room. Hence Burton's stalker joke.

The way Dunmire sees it, *Star Trek* is a morality play as good as anything that Shakespeare or the Greeks wrote, but the appeal to a longtime military man is obvious. It's a leadership fantasy. The Starship *Enterprise* is out there on its own, fighting challenges improvisationally, unencumbered by bureaucracy or sluggish agencies. When the captain decides action is called for, action is what he gets. He doesn't wait for appropriations, or write policy reports, or dance around political correctness. Dunmire's favorite episode, a two-parter from the fifth incarnation of the *Star Trek* series called "A Mirror, Darkly," is revealing. It covers a hot-headed, over-ambitious mutineer who conjures up a wild scheme to save the empire that others have failed to protect. In the episode's climactic moment, the new commander says, "I've been a soldier all my life, and I will not stand by and let these people destroy an empire that has endured for centuries!" He gets down on one knee and continues: "I ask you, all of you—join me."

"My job is to make boring things fun," Dunmire once told me. "Corrosion's not a sexy topic." About the second half he was right, but whether or not *Star Trek* has sex appeal beyond the military is an open question. One can only contemplate a body in a Starfleet uniform for so long. "You pull 'em in," he said. "You gotta pull 'em in. When we're finished, it'll be nifty." In the words of Greg Riddick, Dunmire's consigliere: "They recognize that guy. They're gonna sit up and watch it." Stacey Cook justified the LeVar

Burton approach more squarely. "His delivery is what we want," she said. "It's what appeals to people. It's like, wow, someone talking about technical corrosion, and not just schooling them. It's an extremely soft sell." Indeed, when Burton narrates, he sounds fascinated without sounding erudite, engaged but not obsessed.

Beside the trailer in Florida, I asked Burton, "What do you tell your friends in LA?"

"I don't tell them," he said. "No, I say I'm doing some work with the Department of Defense. Then, when their eyes glaze over, I move on." Woe to West Coast wonks. After a pause, he added, "There was only supposed to be one video." Earlier, while a PA was cleaning a smudge off of a Plexiglas screen, before Burton had even begun shooting the fifth corrosion video, Dunmire asked him to do number six. The conversation went like this:

DUNMIRE: "You know about LeVar Six?"

BURTON: "Six?"

DUNMIRE: "Cook bring it up to you?"

BURTON: "I'm retiring."

DUNMIRE: "Yeah, six."

BURTON: "Five, I'm out."

DUNMIRE: "Yeah, six."

BURTON: "Okay, six. That has a nice ring to it. And I'm out."

DUNMIRE: "Yeah, six!"

BURTON: "I'm game. So this really is paying off, Dan?"

DUNMIRE: "Yeah."

BURTON: "That's great. It really is."

DUNMIRE: "I'm excited."

BURTON: "As am I."

After thinking about it for a few minutes, Burton said, "This is part of my legacy, Dan."

I asked Burton, "Do you do other work like this?" He responded immediately: "There's nothing like working with Dan."

Dunmire has a reputation for being crazy, scattered, and bumbling. He doesn't know his office phone number or address. He wears loud outfits— Penguins shirts, Pirates shirts, a Steelers turtleneck, a Steelers sweater, Pittsburgh-themed ties, layered articles that many would wear only ironically. His Department of Defense ID hangs on a Steelers lanyard. He used to drive a Chrysler PT Cruiser painted like a giant Steelers football helmet. He calls himself a buffoon. People invariably call him a character. In social situations, he lingers a bit too long, awkwardly, as if unaware. Talking to three staff, including the director, of a museum in Orlando, he stood in a hallway with one leg forward, his torso angled and canted forward, both arms out, one finger pointed, the other fist clenched. He held his head sideways, with his evocative eyebrows cocked, his forehead the closest thing to those around him, the pupils of his beady eyes peeking out from above nonexistent glasses. It's a worrisome, piercing posture with a hint of the lunacy you see in the homeless or deranged. Had he been playing charades, I'd have guessed that he was conducting an orchestra, or doing a hippie dance in San Francisco's Golden Gate Park. The museum staff, meanwhile, beheld Dunmire with feet planted firmly, steadily, hands clasped or in pockets, their upper bodies immobile. Only their necks moved slightly, as did their facial features. An hour later, when he bumped into a visitor dressed up as Captain Kirk (it was, after all, Halloween), Dunmire stood up straight, gathered himself into a proper posture, and told him, "Good work." He addressed the costumed man as "sir." Shortly thereafter, he joked that maybe he ate too much lead paint as a kid.

When he speaks, Dunmire's voice is hoarse and scratchy. His words, laden with Pentagon jargon like "DepSecDef" (Deputy Secretary of Defense), come out fast and slurred, always with emphasis. That's the other thing people invariably cite in Dunmire: passion. Talking to him is sometimes like getting berated by a coach—he's always fired up. He's often so excited that what he says is rambling and discursive to the point of annoyance, such that more than once while taking notes I wrote, "Dan is a terrible storyteller." If you have an agenda, it's frustrating. If you want specifics—like, say, the cost of producing a LeVar video—he'll make you bang your head on the wall. If you just want a story, he'll get your attention. He's a whiz at improvising. At Corrosion Forum XXXI, in the middle

of a PowerPoint presentation clogged with numbers, acronyms, and procedures, Dunmire interrupted, took the microphone, pointed to the screen,
and said, "Why do I need any charts? I can just talk about this stuff," and
winged it with a story. Leaning on the podium, he scratched his head,
looked down at the floor or up at nothing in particular, and stole the show.
Nobody could accuse him of lacking dedication or commitment.

Aside from all things Pittsburgh and *Star Trek*, he's got one more ob-
session: he doesn't like goats. Examining corrosion on radio antennas in
Kauai, Hawaii, once, he learned that invasive goats were eating vegetation
so quickly and thoroughly that it was leading to rapid erosion. The foundations of government structures were threatened. "We gotta stop these
goats!" Dunmire said. "We gotta kill 'em!" One colleague got him a Navy
shirt that says, "fear the goat." Others regularly write proposals for getting
rid of goats. Mention the word *goat* and Dunmire will lock eyes, tilt his
head forward, and give you that intense look.

Back inside, Burton held the chunk of iron while saying, "When we
first dig it up it looks like this: a lump of iron ore." The chunk was smaller
and less cumbersome than before, thanks to Kinzie, Dunmire's chief
technical engineer, who broke it in half. Four decades of corrosion work, a
handful of patents, a degree in chemical engineering—and this is how he
helped. Otherwise Kinzie sat quietly in the corner, waiting for technical
issues to arise. One could be forgiven for getting the impression that what
turns Dunmire on is what turns the rest of his team off. He's the social animal, and the rest are the technical guys. They're also his safety net.

When Michael Wynne picked Dunmire to run the corrosion office, he
knew what he was doing. He wanted a program manager, a facilitator. He
recognized that a geek like Data would make a terrible captain. Wynne
wanted someone who could wrangle engineers. But engineers didn't see
the wisdom at first. Rich Hays, now Dunmire's deputy, ran the corrosion
lab at the Naval Surface Warfare Center in Carderock, Maryland. From
Virginia Tech, Hays had undergrad and master's degrees in materials engineering, and he had worked on corrosion-caused cracks in submarine
propellers and on corrosion in expeditionary fighting vehicles (EFVs) belonging to the marine corps. His first impression of Dunmire was "Who is
this joker?" While Dunmire presented, he laughed openly. He thought the
man unqualified.

Hays—tall, skinny, and bespectacled—is precise, direct, linear, and serious. He recalled, "I was still in my mind-set of solving corrosion as a technical issue. I am converted to a large degree. The reason I'm here is to learn how the hell he did it when I couldn't." He said, "We were gonna invent the next best thing. Some technology. Some new coating. We learned quickly that's not the way to go." Now, of Dunmire, he said, "Dan is a visionary. He's the opposite of me." He added, "Dan's really good at starting stuff. I'm here to institutionalize it, get it done. If the office goes away, it'll be like we were never here. So that's why I'm excited about policies." In other words, he's part of a team that keeps Dunmire in line and on track. He resisted Dunmire for years and finally yielded. Now he jokes about planting a GPS tracker on his boss.

Hays wasn't the first to join Dunmire's team—he wasn't one of the rusketeers. The first was Larry Lee, a quiet, diligent, polite air force colonel (now retired) born in the Philippines. After studying chemical engineering and joining the air force in 1977, he spent twenty years keeping aircraft flying. Inspecting and designing fighter jet engines, he rose from an enlisted airman basic to colonel. In the summer of 2001, he landed at the Pentagon, in the Office of Acquisition, Technology, and Logistics, near Dunmire. A year and a half later, Dunmire brought on Lee as his deputy director. Seven years later, when Dunmire finally swayed Hays, Lee became Dunmire's chief of staff.

Dick Kinzie, of rock-breaking skill, was also there from the beginning, as Dunmire's second hire. He had four decades of corrosion work, half of that as a materials engineer in the air force's Corrosion Prevention and Control Office. He worked on the design of new aircraft, the upkeep of old aircraft, and assessing the cost of corrosion. By the time Dunmire hired him, he had risen to deputy office chief and retired. Unlike most, he knew his corrosion counterparts in the army, navy, coast guard, and NASA, which made Dunmire's life easier. In the early years of the Corrosion Prevention Office, for all who had the same impression that Hays did, there was Kinzie, allaying their fears with his gentle southern warmth and calmness. He too had his doubts about Dunmire's chances of success. He's become Dunmire's insurance policy. Chief technical engineer he remains.

The men who comprise Dunmire's core team are nothing like him. Each is an opposite: quiet, linear, calm, composed, less obsessed, closer to "normal," technically proficient. When Dunmire stumbles over tiny details—whether an acronym is C&O or O&C, whether lines on a chart point in the right direction—and digs a hole for himself, his sidekicks correct him. They reassure others that the agenda, the program, is not being winged on the fly, that it has bounds, that it can work in this galaxy at this star date. Dunmire won't take notes or follow a schedule, but they will. Dunmire will banter, and they'll wait. As Larry Lee put it, he and Hays will be in the kitchen, Kinzie will be in the garage, and Dunmire will be, well, roaming around somewhere outside. Yet they get along. In ten years, not one has quit. Nor has anyone been fired. "My team is not bullshit," Dunmire told me once. "I may be bullshit, but my team isn't." He has called them geniuses. Often they wear matching CPO shirts, in orange, khaki, or blue, but it's obvious which one contains Dunmire. At Corrosion Forum XXXI, attendees called Dunmire eccentric, unconventional, and non-Pentagony, but they also called him charismatic, guileless, spellbinding, effective, and great. One person put it succinctly: "All of us worked in corrosion forever, and we never got anywhere. Then Dan came along. He's funny, he's got energy. He's colorful."

It sounded like a dive bomber, but it was just a prop plane flying overhead. Burton, working from the top of page 11, stopped reading while everyone waited for the audio guy to give the all clear.

Dunmire said, "Hold on. This won't fly. What the DOD did was create a culture change. We can't say *needed* a culture change. That won't pass security review." He explained this to Cook—"Not a culture of change, a culture change"—and to the teleprompter editor, and to Burton, who laughed. Burton repeated what Dunmire said, skeptical of Dunmire's insistence on minutiae. It seemed that Burton had begun to tire of the DOD manner, to suspect that Washington was hopelessly bureaucratic and gargantuan and nitpicky. Dunmire found another issue. He said, "Technically, *under secretary* is two words, capitalized." Then, to nobody in particular, he said, "I'm the chosen one." Burton didn't respond to the second comment. He just said, "I'll make it two words in my mind." What he meant was: Who the fuck cares?

Burton, though, believes in Dunmire's program, which is why he's continued to shoot the rust videos, and why he bills Dunmire at a reduced rate. It's also why, at Dunmire's 2011 corrosion conference in Palm Springs, California, he praised Dunmire in a keynote speech, comparing him in spirit and authenticity to the greatest men he had known: Alex Haley, Gene Roddenberry, and Fred Rogers. Haley wrote *Roots*, the 1977 miniseries that starred Burton; Roddenberry was the creator of *Star Trek*; and Rogers created a much-loved PBS children's program, *Mister Rogers' Neighborhood*. Burton called Dunmire's commitment to the greater good genuine, honest, and boundless. He called him a force of nature. Burton also admitted he'd become a rust evangelist because of Dunmire.

Assured that the script was now correct, Burton resumed his dramatic reading. "All they needed now," he said, "was a plan." Dunmire raised both hands, flexed, and said, "Oh, yeah. I love the way you say that."

⁓

Dunmire comes from a long line of headstrong military men. His tenth-or-so-generation great-grandfather, a German farmer named Johann George Dormeyer, settled in Allegheny County and fought in the Revolutionary War. His five sons all worked the five-hundred-acre family farm in Elderton, northeast of Pittsburgh. In 1810 a census official changed the family name to Dunmire. For seven generations since then, every Dunmire but one, a cabinetmaker, was a farmer, and when circumstances required, a soldier. One Dunmire or another fought in the Civil War, World War I, World War II. None died in battle. None earned a college degree. Dunmire's great-grandfather Samson Dunmire died after getting kicked by a cow. Samson's son sold the family farm a decade before Dan Dunmire was born.

The baby in the family (Dunmire has two older sisters), Dunmire took after his father, a steelworker at U.S. Steel's Ohio Works in Youngstown, Ohio. "Do it right the first time, be responsible, take care of business"— that was his dad's philosophy. His dad kept inconsistent hours, came home dirty, and didn't blame anybody for mishaps. No engineer (he didn't finish college), he nonetheless knew corrosion.

From the time that Dan was two years old, he knew he would join the army. He waited until he was seventeen, whereupon he signed up for the

ROTC. At Kent State, he wore his uniform with pride. This, in the years following the 1970 massacre in which four students protesting the US invasion of Cambodia were gunned down by members of the Ohio National Guard. "I was very unpopular," Dunmire recalled. "I said, 'God bless America.'" Yet he was also more liberal than his father. A constitutionalist, he saw the government as serving a role in private life, particularly as it related to Pittsburgh. In 1970, at the opening game at the Pirates' new ballpark, Three Rivers Stadium, Dunmire argued with his father about public/private partnerships. Dunmire the elder thought the stadium ought to have been funded privately. Dunmire the younger thought the government had a role in supporting the city's culture. To this day, he thinks that Pittsburgh wouldn't have a concert venue, PNC Park (the Pirates' newest home), Heinz Field (where the Steelers play), or be considered a major city if not for public investment.

In 1974 Dunmire volunteered to go to Vietnam. He was sent to Germany instead, and spent three years in the US Army's VII Corps, first as a lieutenant, and then as a platoon leader. His task was to create denial barriers, anything to slow Soviet forces down should they decide to head west. He pretty much spent three years blowing up roads and bridges. Now, as the defender of military infrastructure, it's almost like he's atoning for those acts.

Returning to the States, Dunmire landed in Tampa, and began working in food service. He spent the final years of Carter's presidency as a shift manager at the Busch Gardens theme park, and the first year of Reagan's presidency in a Marriott. The hotel chain sent him to Birmingham, Alabama, where, on top of work, he began earning a master's degree in public administration. He met his wife. At a Veterans Day dinner, he also met Hyman Rickover, his longtime hero. The admiral was old and frail, and Dunmire probably shook his hand a bit too enthusiastically while thanking him for his contributions. Dunmire loved him because he was a master of bulldozing through bureaucracy, was well rounded, had a sharp team of subordinates, and—with his four stars—refused to kiss ass. For the last trait, after becoming the longest-serving man in US military history, Rickover was fired. He died four years later.

On April Fool's Day, in 1982, Dunmire bought an establishment on the edge of Alabama's campus called Doogie's Hot Dogs. He changed the name to Dan's Breakfast and Lunch, and ran it for two years. He remember-bers his duties as owner and chief bottle washer. For twenty-six months, Monday through Friday, he got up hours before dawn so that he could open the place at six. Dunmire took classes at night. On weekends, he served in the reserves. He sold the restaurant when he was selected for the Office of the Secretary of Defense's Presidential Management Internship program. That's how he first got to the Pentagon. It was 1984.

It's not like Dunmire had gotten rich selling hot dogs and Coke; in Washington, he and his wife rented a second-story apartment in the historic Anacostia neighborhood. After the internship in the Office of the Secretary of Defense, he spent another year there working as an analyst, looking at personnel compensation and readiness. Then Dunmire became an analyst in the AT&L office. He moved to Temple Hills, Maryland, just southeast of DC, between the Pentagon and Andrews Air Force Base. Eventually he found a foreclosed home—a bargain buy—on a half acre in Stafford, Virginia. He had three kids.

In the mid-1990s, during the Clinton presidency, Dunmire applied for the Undersecretary's Best Commercial Practices fellowship program, which would put him in private industry for two years. For his choices, he picked food-service companies, which were familiar and presented no conflicts of interest. He listed Heinz, Pepsi, Coca-Cola, the Reynolds Metals Company, and Anheuser-Busch as potential employers. His preference by far, though, was Heinz, in Pittsburgh. Dunmire prepared thoroughly. When interviewed, he cited profit and loss figures, quoted authors, and mentioned numbers the interviewers had never heard. The interviewers told him they'd call him in two weeks. At home, there was a message on the answering machine. It said, "Can you come two weeks early?"

He spent the next two years in Pittsburgh and in various Heinz facilities around the country studying how the company did business, and in particular how it bought brown paper. Heinz bought $20 million of paper a year, for boxes, trays, and packaging. He went to paper school at Weyerhaeuser, outside of Chicago. For four days, at one of the country's largest paper-products companies, he immersed himself in the minutiae of forty-pound brown paper. He learned about warp, miscuts, corrugation; he made himself knowledgeable. When Heinz closed a plant in Tracy,

California, Dunmire helped integrate the extra ketchup line into a plant in Fremont, Ohio. He made sure that the boxes would run through machines. For $250,000, Dunmire said he could get the line configured and running, and in a year, he told a vice president, the company would recoup the cost. At the largest ketchup plant in the world, his plan worked, and the VP was impressed. Dunmire still calls that line his line. At the end of his internship, Heinz offered him a job as a senior purchasing agent. He was intrigued—"I could have made it in industry after all," he said—but, by then in his forties, he figured it was too late to make such a big switch. The military was calling. Heinz, Dunmire says, made him aggressive, gave him a critical eye, and changed his perspective on business. It also made him think of work as more of an adventure than a job.

Back at the Pentagon, Dunmire worked on Bush's transition team. When everybody else was on vacation, Dunmire wrote a three-hundred-page report. The DOD's general counsel noticed. When new staff were hired, and they had yet to be cleared by the FBI and confirmed by the Senate, the Pentagon had Dunmire escort them around. It is unclear whether he did this wide eyed and leaning forward. In this way, he met Michael Wynne.

In Dunmire's world, two major events took place in September 2001. After the first, air force colonel Al Evans, the military assistant to Undersecretary Wynne, asked Dunmire a not entirely hypothetical question. If they had to leave the Pentagon, he asked, what key acquisition documents would they need? Dunmire emailed a dozen documents to Evans thirty minutes later. That same month, NACE published its cost-of-corrosion study.

Wynne sent Dunmire to the Raven Rock Mountain Complex, Pennsylvania's ultrasecure "underground Pentagon." He spent six months there. When he returned, he went back to acquisition resources and analysis, looking at chemical demilitarization, guided munitions, weapons systems. On top of the usual matters, he met with Maren Leed, from the Senate Armed Services Committee, to work out this corrosion thing on Senator Akaka's plate. His boss was the number three man at the Pentagon, below the secretary of defense and the deputy secretary of defense. (Protocol has it that the four undersecretaries of defense—for AT&L, policy, the budget, and personnel and readiness—follow that pecking order.) Then Congress passed a bill, Bush signed it into law, and Wynne called his staff to a meeting.

Six years and many GAO audits later, Congress rewrote the law regarding the corrosion executive. Like Wynne, successive undersecretaries had decided that they too could fill the role. Congress wanted someone dedicated to the job. To a new director, the authorities of the old executive went. Dunmire, by then a special assistant, was the obvious pick.

⌁

"You all ready for talent?" Cook asked. "Okay, bringing in talent." It was hotter in the studio, and over the next two and a half hours, Burton had the sweat wiped off his forehead at least thirteen times, just about every ten minutes. Dunmire, meanwhile, grew visibly tired. His ability to sit up straight waned. He yawned, rested an elbow on the table, rubbed his temples with the thumb and index finger of his right hand, drank a can of Coke, yawned again, stretched, groaned, rested a hand on his knee, took a deep breath and exhaled, yawned loudly, and then stretched. Shifting positions, he knocked over a coffee cup. I asked him if he planned on taking a nap. "Are you crazy?" His eyes widened. "I didn't drive all the way from Pittsburgh to take a nap!"

He continued making minor changes to the script so that the video would survive intact. He added the phrase "environmental severity index" to a sentence about the most corrosive environments in the country, and changed the word *built* to *completed*. He corrected *2002* to *2003*, and insisted that military academies—the US Military Academy, the US Naval Academy, the US Air Force Academy—be listed in order of creation. When he questioned whether three hundred million dollars should be referred to as hundreds of millions of dollars, Cook told him he was just being picky. When he suggested that "we've" become "I've," Cook told him not to worry about it.

Reading from page 18, Burton explained, "Before the corrosion program started, each of the different services had its own rules for dealing with corrosion, and that usually meant wait till it corrodes, and then repair it." Then he ran through some of the projects that Dunmire's office has funded. He started with an example of an asset that the navy couldn't afford to let keep corroding: Hawaii's Red Hill pipeline, a thirty-two-inch line running three miles between twenty massive buried fuel tanks and Pearl Harbor. The pipeline was built in 1942 and kept a secret until 2005, when Dunmire decided it was time to "pig" the line. It had never been inspected internally. Burton roughly explained pigging—"a process that sends sensors down the pipe interiors to check for problems"—and then moved on. What he didn't mention was that Dunmire threw $1 million at the project, his office's first. Dunmire saw it as low-hanging fruit. Nor did Burton mention that the navy now owns a custom ultrasonic pig made by Vetco. Unmentioned also was that just after the inspection, a day or two before Christmas, the line sprung a leak, while crews were there to deal with it. Had they not been present, the leak could have tainted Honolulu's water supply and inflicted $1 billion worth of damage. Because it was contained, the leak didn't make national news. But Senator Akaka found out and later invited Dunmire to sit beside him at his office in the Hart Senate Office Building. Akaka called him Dormeyer.

Since 2006, when Dunmire got his first *true* budget—$15 million, approved by the White House—he has spent roughly two-thirds of the money on projects: half on weapons projects and half on facilities projects. After metrics (studying the cost of corrosion), weapons and facilities projects comprise the second and third arenas on which Dunmire's ten-person staff focuses. Red Hill pipeline notwithstanding, the facilities projects tend to get overlooked because they address the same problems everyone else has: rusty pipes, pumps, roofs, tanks, and so on. The weapons projects, for the benefit of helicopters, Patriot missiles, and aircraft carriers, are far sexier.

A page later, Burton mentioned technical collaboration with universities and government labs, and the development of a corrosion-sensing paint at a University of Southern Mississippi laboratory. What he didn't mention is that from 2005 to 2013, of the 236 weapons projects that Dunmire's office has thrown $165 million at, roughly a third of them have been in pursuit of the perfect paint. The Corrosion Prevention Office has funded the development of coatings for aircraft, decks, fire systems, jet fuel tanks, water tanks, air-conditioning coils, pump impellers, vehicle underbodies, bilges, magnesium parts, and cold environments. They're single coats or multiple coats, primers or topcoats, designed to cure quickly or at high temperatures or low temperatures, for spraying or rolling or powder coating or depositing by laser. Some are magnesium rich, or zinc rich, or vinyl based, or epoxy based, or nickel titanium based, or specifically chrome free. Some are fluorescent, stealth, sticky, thick, long-lasting, flexible, fire-resistant, chip-resistant, thermally insulating, or nonskid. The office has put more than $3 million toward paints that are self-priming, self-inspecting, self-cleaning, or self-healing. The office has also spent just under $1 million developing peel-and-stick patches for the marine corps, so that soldiers can quickly repair said coatings in the field. Dunmire calls coatings "the first line of defense" and says, "Sometimes you don't have any other solution."

Many of the paints spent years beside the warm, salty water of southern Florida, where scientists at the Naval Research Lab examined their fortitude much in the way that a kid might study milk by leaving a carton in a refrigerator for months. Stuck to playing-card-sized pieces of metal, the paints endured atmospheric exposure rivaled nowhere in America but on Cape Canaveral (where many were also tested, with help from NASA). On rows and rows of racks, the painted coupons blossomed with rust, exfoliated with rust, bubbled with rust. Many faded from gray to tan or blue or even pink. The navy does not go for paints that turn pink. To those paints that the navy has gone for, the staff of the NRL composed a sort of military paean:

> The purpose of the hull is to protect the paint.
> The purpose of the reactor is to drive the paint around.
> The purpose of the SUBSAFE program is to ensure the paint comes to
>   the surface and will not be lost.
> The purpose of the cathodic protection system is to back up the paint.
> The purpose of the weapons is to defend the paint.
> The purpose of the Special Hull Treatment is to protect the paint.
> The purpose of the Vertical Launch System is to destroy those who
>   would do the paint harm.

The military's best paints used to incorporate hexavalent chromium, the carcinogenic stuff at the heart of the 2000 movie *Erin Brockovich*. It's phased that out, and Dunmire's office is now behind an analysis of alternatives: a "program management guide to hexavalent chromium." Seeing a mockup of the guide at Corrosion Forum XXXI, Dunmire called it phenomenal. He gets excited about paint. With a good paint, properly applied, a military asset may live long and prosper.

⟶

Since 2006, the budget suggested by the White House each year has fallen steadily. No matter the size of the presidential budget, Dunmire insists that his office is adequately funded. "We do what we can with what we have," he said. "If you can't take it, get out." Like his father, Dunmire doesn't complain. He has fans and enemies alike, but he wants his results to stand on their own.

While the executive appropriations decreased, the number of projects his office supported slightly increased. At the same time, Dunmire's priorities have changed. In the initial years of the office, the projects tended to be either low-hanging fruit or focused on corrosion sensing: sensors for cathodic protection systems, for finding leaks or cracks, for finding rust on ships, for detecting environmental exposure, for seeing rust in fuel and ballast tanks. Once the easy targets had been hit, the office devoted resources to bigger, less glamorous headaches: inspecting corroded guy wires on antenna towers or tackling rust at Guam's Kilo Wharf. Lately, the office has focused on material selection and corrosion inhibitors and on technology applicable to all branches: composite bridges, concrete docks with noncorroding rebar. Much of it is basic stuff: aircraft washdown and rinse systems, dehumidifiers with long hoses, mildew-removal kits; aircraft covers and shelters. Some run tabs into the millions, but others—like laser powder deposition to repair seals on jet engines—have gone for as little as $30,000.

All the fancy paints in the world don't do any good if the painter doesn't know what he's doing, though. One could be forgiven for assuming that paint is a simple business and that minimal education is sufficient for its application—but neither is the case. At shipyards and hangars, fewer than half of the military's painters possess high school diplomas. Many can't add or multiply. When mixing paint, they use a ladle of this and a ladle of that, to avoid math. Assigning junior members to painting has virtually ensured poor results. Failing to mandate rigid specifications for the military's vast painting jobs has guaranteed them. The only thing more boring than painting a navy ship is coming up with the standards by which it must be done. Toward this end—the fourth arena of Dunmire's office—the corrosion team has devoted considerable energy, reviewing thousands of specs that govern, among other things, humidity, surface preparation, application, and thickness. Dunmire saw opportunity in the problem.

Broader and more potent than technical specs are the policies that Dunmire's office is trying to institute. These are Dunmire's fifth arena of focus, and as Chief of Staff Larry Lee puts it, "where you make the big bucks." When the Office of Corrosion Policy and Oversight was established in 2003, Dunmire and his sidekicks took only cursory looks, when lucky, at weapons under development by the military. Now all DOD programs get shuffled through the office, keeping one member of Dunmire's staff more than busy. He looks at plans for the air force's new KC-46 aerial refueling jets, its new F-22s and F-35s, its new long-range radar, its $6 billion Air Force Space Surveillance System, or Space Fence. He looks at the DOD's UHF satellite communications system and the marine corps's successor to the Humvee, the Joint Light Tactical Vehicle. For the navy, he looks at the new tactical jamming system and the new Air and Missile Defense Radar, new combat logistics ships, Poseidon planes, and presidential helicopters. All military weapon requests for proposals (RFPs) now stipulate corrosion prevention and control assessments. Dunmire's office established guidelines for companies submitting new products to the military; in the twenty-first century, companies must address corrosion before proceeding. Dunmire has called this "getting corrosion policy weaved into the fabric of the DOD." Now he wants all federal contracts screened for corrosion. He wants to change the policies governing Federal Acquisition Regulation, such that any contractor taking federal dollars must be overseen by a certified third-party corrosion officer—from NACE, or the Society for Protective Coatings, or some other agency. Dunmire calls this a final pillar and says it will reduce the cost of corrosion by 30 percent.

Burton was down to the penultimate page of the script. He mentioned the sixth arena of Dunmire's focus—training and certification—the purpose of which is to ensure that "there are always people truly qualified to carry on the war against corrosion." Dunmire wants warriors to know their enemy, and certifies that they do. Since 2005, nearly two thousand military men—mostly from the navy—have taken corrosion courses through NACE. It offers a five-day course and an accelerated half-day version. Only a hundred army soldiers have taken either. For painters, the Society for Protective Coatings now offers DOD-tailored courses on painting techniques and skills such as abrasive blasting, airless spraying, coating thickness—that kind of stuff. Three-quarters of the courses are taken by navy sailors; three-quarters of them pass the test at the end. Thanks to Dunmire's efforts, the service academies are incorporating corrosion in their curricula. The United States Air Force Academy was the first to do this; the others are readying to deploy. The Defense Acquisition University is developing online corrosion courses on a far grander scale, for tens of thousands of service members and contractors. As of late 2012, its platform could handle a million users, but fewer than five hundred people had created accounts.

"Perhaps the most innovative contribution of the training and certification team," Burton said encouragingly, "is the University of Akron's bachelor of science in corrosion engineering." It's the first and only corrosion degree in the country, and it was schemed up by Mike Baach, a former corrosion industry executive. Baach had founded and taken public in 1992 a company called Corrpro, and recognized a general unawareness of the potential benefits the rust industry could bring to the American economy. He also recognized a lack of skilled corrosion engineers he could hire. For thirty years, he'd dreamed of a college that offered an undergraduate corrosion engineering degree. When Dunmire came along, he introduced him to Sue Louscher, now the director of Akron's program. Baach calls the undertaking his proudest accomplishment.

Dunmire first got behind the program in 2006. Since 2008, when it was congressionally initiated, $35 million tax dollars have gone its way via the DOD. Doors opened in the fall of 2010; as of 2014, sixty students are enrolled, including a dozen in their final, fifth year. The way Larry Lee sees it, those students are guaranteed good, high-paying jobs as soon as they graduate. Many of the students have already joined a student chapter of NACE, which they call a corrosion squad. At NACE conferences, travelling in uniform, the corrosion squad resembles a track team without the bulging quads. Of the program at the University of Akron, Burton said, "It's training the next generation." Then he gave a funny look and said, "The *Next Generation*. I like the sound of that."

Star Trek jokes aside, Dunmire is dead serious about the program. It's the pinnacle of STEM development, spitting out not just engineers but *corrosion engineers*. Taking full advantage of the educational opportunities, Dunmire has in the works "learning modules" created by "subject matter experts" at the University of Akron, edited by Bruno White studios, and deployed to tens of thousands of soldiers online via the Defense Acquisition University. Describing this plan and its "content," Dunmire and his cohorts employ a great deal of new media jargon, making it hard to discern fantasy from reality.

There's one more STEM program, on Earth, that really excites Dunmire. As part of the office's seventh and final arena—outreach—it funded a small exhibit at the Orlando Science Center called *Corrosion: The Silent Menace*. The exhibit debuted in March 2013, in Science Park, a section of the museum with displays on electricity and magnetism, lights and lasers, gravity, potential energy. The exhibit featured a fourteen-foot trestle bridge, with chipped paint, concrete breaking off, exposed rebar, spalling—a half dozen manifestations of corrosion. It had the sounds of traffic recorded from an actual bridge, but everything else was simulated. The concrete pedestal, the steel I beam, the steel bolts, the rust—all of it was made of plastic and paint. It seemed as silly as making fake dirt, or fake trash, when so much is so plainly available. It was like going to Taco Bell in Tlaxcala. The height of irony was spending $75,000 of our tax money to fake the look of decay, when the DOD is spending 266,000 times as much money to eliminate it. But you can't have museumgoing kids getting crushed by a collapsing beam.

Maybe kids don't care—Dunmire is no child-education specialist. If the exhibit doesn't excite kids, Dunmire figures at least it'll expose them to rust. Daily, the museum plans to teach eighteen kids more about rust in a small classroom. There was even a career kiosk, with a recorded video intro by LeVar Burton. Dunmire wants the exhibit to serve as a template—a launchpad for rust exhibits nationwide. With private funding, he hopes that other museums will make copies for a quarter of the cost. In the meantime, he hopes that the Orlando Science Center, at least in ticket price if not attendance, is more appealing than Universal Studios.

It's LeVar Burton, though, that really puts a smile on Dunmire's face. LeVar Burton talking about rust exemplifies the outreach that is unique to

Dunmire. The videos have wide appeal and captivate civilians, or he hopes they will. He's counting on it. "Darryl wrote it, but it's my script," Dunmire said of LeVar 5. "This is what I wanted. This is mine. This one's mine, all mine. This is it." Dunmire also calls the office his baby.

More than once, Dunmire refused to give me a cost per production of the LeVar videos. He got touchy, said it was impossible to pull out the cost of one movie when they were all commingled and budgeted with other related projects, insisted that he follows federal rules and contracting procedures, said he'd get back to me on a ballpark figure but only guaranteed that they cost less than $300,000 each (most of which went to editing). At the same time, he reminded me that his office is the most audited in DOD and said, "I'm a frugal sonofabitch."

By three thirty, LeVar was almost as tired as Dunmire. He goofed up: said "fuel surprise" instead of "fuel supplies." He said, "I may make it look easy, but this shit's work. There's only so much juice in the battery. I degrade. I rust, Dan." Dunmire laughed. Soon after, in the middle of another take, the jib crane squeaked, so Burton had to take two. He grimaced.

Dunmire flipped to the last page of the script and said, "The big finish." Burton said, "Get your Kleenex box ready, because it's gonna be poignant." Per custom, he read: "Fighting corrosion is a job that's never really finished, because, as we've said before, rust never sleeps."

Lord said, "That's a wrap." Dunmire and Burton hugged, patted each other on the back, and shook hands. Then Burton stepped outside to smoke.

It took Dunmire years after he assumed his throne to discover how little faith others had in him. Neil Thompson, a former NACE president, wondered how he got anything done. "You're not a linear thinker," he once said accusingly. Paul Virmani, the author of the 2001 cost-of-corrosion study, doubted Dunmire's abilities. "You're not an engineer," he said in the same accusatory tone. "How can you possibly do the job if you're not an engineer?" Ric Sylvester, the longtime senior executive in the AT&L Office, told Dunmire that he made others look bad, and also jealous. (Sylvester also told Dunmire that he'd never succeed because he refuses to sacrifice his ideals.) When the Potomac Institute for Policy Studies asked Dunmire to join its peer review board, it performed a background check with current and former officials and servicemen. "We want you to know, you are not very well liked," someone told Dunmire. "But you're respected because you get things done."

Dunmire claims otherwise. He thinks of himself as one of the most disliked guys in the DOD, as unpopular as he was at Kent State. A mutha-fuckin' sheriff he is not. He says it's not respect that he's earned so much as it is fear, the result of his refusal to brownnose. He knows that he's a pain in the ass. When he gets read into a program, he says, others end up un-happy. "We are screwed," they say.

Dunmire doesn't pay it any heed. "I don't care. I do the job. I'm not here to kiss anybody's ass. I stay within the boundaries." He elaborated. "I'm serious, that's why I've survived for ten years. I don't cross the line." Per custom, he fell back on his refrain: "I love the warrior."

Dunmire insists he's no maverick ("I can be calm and rational; I just enjoy being crazy") or jerk ("I'm not an asshole; I'm just passionate about what I do") or corrupted official. For all his weirdness, he's never mean spirited or abrasive. He insists he's following his own compass ("I know what I need to do next") and that when he goes to the edge, he informs his command of his plans. He says he's taking risks that need to be taken and that it's his own tail on the line ("I'd rather be lucky than good"). Corro-sion, he figures, demands a willingness to gamble ("Nothing gets done by playing it safe") as much as it requires innovation ("We're fighting the sec-ond law of thermodynamics, so you have to do interesting things"). Rust, he says, is tougher than paper. Often it's difficult to justify ("It takes mil-lions to save billions"). He hits his old refrain, that he loves what he does, and underlying it all is a love of the warrior.

Every time we talked, Dunmire emphasized that he's just a cog in a wheel. He always deflected attention elsewhere; said the administration deserves any credit. He's just a bureaucrat, he says, proud to be a bureau-crat. "All I do is facilitate." More than once, he suggested I not write too much about him. "Shorter is better, you know." He advised I make his profile pithy. Occasionally, he expressed astonishment that he ever made it into the Office of the Secretary of Defense, which he calls Nirvana. He's grateful that his position is not at the status of political appointee. "You think I would ever get confirmed by the Senate?" he asked rhetorically. "You think the White House would nominate me?" Of his presence in LeVar 5, he said, "The only reason I put myself in this movie was because you're talking about the creation of the department. I didn't wanna be in the movie. It's like I have a big ego. But I want it timeless. I want it to have long legs. I want it to be history."

Presented with a podium, though, Dunmire always climbs up and revels in the position. In the DOD's maze of strong personalities and giant egos, he competes in his own way. Tony Stampone, a logistician in the Office of the Assistant Deputy Undersecretary of Defense for Materiel Readiness, has known Dunmire since 1989, when both were GS-13s. Dunmire and Stampone hail from opposite sides of Pennsylvania and worked on opposite sides of the Pentagon. Dunmire, Stampone recalled, always seemed a little bit on the edge. "I never saw it until the corrosion stuff," he said. "Dan took a shoebox and a small budget and through sheer force of personality made what he could." Stampone said, "I'm a logistician. I pay for this stuff in the long run. Nobody gives a shit what happens. So we care about corrosion. That's what Dan's been preaching. No one had heard of corrosion before. No one cared. Dan got the word on the radar. If you can do that in the Pentagon, that's half the battle."

Rich Hays, Dunmire's deputy, said that as a result of Dunmire's work, where program managers once didn't want to get called out on *60 Minutes* for putting lead paint on a ship, now they don't want to get called out on *60 Minutes* for designing a ship that lasted ten years instead of twenty.

When Alan Moghissi, a respected scientist, professor, and author who's advised the CDC, the EPA, the National Science Foundation, and the Department of Energy, first met Dunmire in Alexandria, Virginia, he liked what he saw. Moghissi has a doctorate in physical chemistry. His son Oliver has served as the president of NACE. He knows corrosion well. "Dan is perfect," he said. Educated in Switzerland and Germany, he retains a strong German accent. "Very often, engineers don't see the forest for the trees. But Dan, he's seeing the forest," Moghissi said. "The Ford CEO is not an automotive engineer. DuPont's CEO is not a chemist." Dunmire, he said, knows that engineers don't consider him a peer, but he also knows what he doesn't know. Moghissi said it was easy to figure out Dunmire.

Halfway through his drive south, in the middle of sideways South Carolina rain, Dunmire called me and said, "I'm sixty years old. I'm crazy! But I'll do anything for the warrior. Anything for the warrior." His devotion he sees as sacrificial service. A dozen times, he expressed his devotion, as if I were a priest. "I care about the warrior. I love the warrior. I do this job for one reason." For a decade running the Office of Corrosion Policy and Oversight, he's put up with the burdens of government contracting on behalf of the life of the warrior. At the start of that decade, his wife said to him, "You're going to die a lonely man because all you care about is the warrior." Acknowledging as much, he's called her a saint. "Without the warrior, we don't have a country," he said. "Any discomfort that comes up, when you think about people giving up life and limb for this country, you think I give a shit what I have to go through? I'll put up with any bureaucratic bullshit that anybody puts in front of me. I love the United States of America. Go ahead and challenge me. I ain't fucking leaving."

On a March evening in 2008, Dunmire almost left. Headed home on I-95 in his PT Cruiser—the two-ton helmet—he fell asleep and hit an exit pole and three trees. He'd been going over sixty. He never braked. An emergency crew cut him out of the car and took him to a hospital. He had a severe concussion and shattered ribs. Dunmire spent four days in the hospital, and when he woke up, doctors told him that he'd actually died.

Two years later, on his way to a Steelers game, he slipped on a sidewalk and broke an ankle. He missed the game. Colleagues aren't sure which was worse: the injury or the result. Since then, Dunmire has put on weight and stopped scrambling around outside like a Dormeyer. Velcro Dr. Scholl's have become part of his uniform. Exhaustion seems to have latched onto him. At Corrosion Forum XXXI, Dunmire was so tired he didn't stand up when people greeted him, and took hugs sitting down. I saw him rub his temples, slouch, shift positions, hunch his shoulders, yawn, wipe his face, twist, and rub his eyes. "I'm very enthusiastic on the inside," he said. On the outside, he looked like a rusty asset.

⌒⌒

The LeVar videos—which Dunmire calls "solid training videos"—make Dunmire jump up and down like a kid at Disney World but don't have the same effect on the senior Pentagon officials. *Reading Rainbow* promoted

childhood literacy. The corrosion videos promote . . . raising adults' awareness. Shortly after LeVar 4 was produced, the principal deputy undersecretary of defense for AT&L told Dunmire's deputy, "Nobody watches your
videos." At Corrosion Forum XXXI, Dave Pearson, a professor of engineering management at the Defense Acquisition University, said his bosses
hadn't bought off on the videos yet.

For all of his outreach efforts, Dunmire's name rarely pops up in non-
military media. *Waste & Recycling News* picked up an army press release
about a thermoplastic bridge that Dunmire's office tested at Fort Bragg.
BusinessWeek published eight paragraphs about the DOD's massive rust
headache and gave two short sentences to Dunmire. The author pointed
out that with the money the military spends on rust in one year, it could
buy two new aircraft carriers or four dozen fighter jets. He called Dunmire
optimistic and left it at that.

Though the budget suggested by the White House has decreased
steadily, to almost half of what it was in 2006, Congressional add-ons—
earmarked for specific projects or not designated at all—have flowed toward Dunmire's office in more than equal and opposite fashion. The office
got an extra $13 million in 2008, as much in 2009, and slightly more in
2010. Congress sent more than $30 million in additional funds to Dun-
mire's office in 2011 and in 2012, and nearly as much in 2013. As a result,
the overall budget of the Office of Corrosion Policy and Oversight is now
more than double what it was in the prerecession years of 2006 and 2007.
Not surprisingly, Dunmire has allies in the House and Senate Armed Services and Defense Appropriations Committees. In the days of restrained
budgets, furloughs, sequestration, and conference cutting, Dunmire's tiny
DOD office—among thousands—is one of the few that is growing. It's
growing because Dunmire gets results.

Pigging the Red Hill pipeline may have been a great success and
pleased Senator Daniel Akaka, but it's a little antenna gasket that Dunmire
points to as a phenomenally successful early project. The US Coast Guard
was the first to complain about corrosion where antennas protruded from
Dolphin helicopters. Dunmire directed a few million dollars toward the
trial of conductive gaskets made by a company called Av-DEC (Aviation
Devices & Electronic Components). The coast guard tried them in 2005
and reported a twofold return on investment. Then the other branches

signed up. The air force used them on C-130 Hercules military transport planes. The army put them on Apache helicopters. The navy used them on Prowler and Hornet planes, as well as Seahawk helicopters, and saved twenty thousand hours of maintenance per year. In 2007 the return on investment was 175-fold. Of the Av-DEC gasket project, Larry Lee called it "the most wonderful story you'll find on a successful project." Dunmire called the project a "home run golden-child money saver." The return on investment was high, but not unusual for a project out of his office.

Dunmire knows he's doing well because he rarely sees the secretary of defense. Like a schoolkid, he aspires to stay out of the principal's office. But he's also doing well because his projects almost inevitably yield incredible returns on investment. What he'd been preaching for so long—that prevention was far smarter than repair—has panned out. It's no longer prophecy. Water rinses and covers for helicopters have returned, respectively, elevenfold and twelvefold their initial investments. Sealants for cable connectors on Patriot missiles score 12:1. Dehumidifiers for the same missiles score the same. Induction heaters for repairing the nonskid on flight decks score 45:1. Paints tend to fare even better. The navy currently spends more than $100 million a year on deck coverings—aka nonskid. A spray-on nonskid funded by Dunmire's office scored 33:1. Fast-cure stealth coatings scored 52:1. A quasi-crystalline coating that's as slippery as Teflon and as hard as stainless steel scored 126:1. Overall, the GAO has predicted an average return on investment for projects in Dunmire's office of 50:1. In other words, over the last decade, Dunmire's office has saved billions.

The day after the shoot, Dunmire woke up late and looked and sounded worse than ever. He'd failed to recover from the all-night drive, let alone the draining video shoot—and had barely summoned the energy to go out to dinner with Burton and a few of the crew. "The spirit was willing," he said, "but the body was weak."

He opened up USA Today to coverage of the destruction wrought by Hurricane Sandy. He said it was like Pearl Harbor without the firebombs. He worried about damaged military equipment, and, seeing a photo of a flooded New York City subway, said there was saltwater in all kinds of undiscovered places—that we'd find out about it years hence. "I just hope they consult us, include us in the contracts," he said. He carried his things into his Ford Excursion—the V10 behemoth, he called it—and loaded them in the trunk. On his way to Florida, his trunk had been full of *Star Trek* paraphernalia: a McCoy model, a set of *Star Trek* Pez dispensers, a door chime à la Starship *Enterprise*, a big blue foam "live long and prosper" hand, and ten more figurines and scenes and toys from the show. He'd given them all to Stacey Cook for Christmas. Now he had a large 3-D television, for corrosion education purposes, in their place.

Dunmire dropped me off at the Orlando airport. The last I saw of the corrosion ambassador was his license plate, PGH 57—for Pittsburgh and Heinz—as he headed north, back into the maw of a natural disaster.

7

WHERE THE STREETS ARE
PAVED WITH ZINC

A chunk of the rust market belongs to the galvanizing industry. To get a sense of that chunk, I met with Phil Rahrig, the executive director of the American Galvanizers Association. The AGA's office is in an inelegant brick building in suburban Denver, alongside a dentist, orthodontist, chiropractor, and CPA. Inside, it's reminiscent of a small environmental nonprofit but without the young, inspired people. When I arrived, on a hot August morning, Rahrig was on a conference call, so I admired the showiest wall of the office while I waited. It held framed shots of various award-winning galvanized structures: a New York City bus shelter, a natural-gas unloading facility in Texas, an air-traffic control tower in Memphis, a chair lift at Utah's Park City, the Indianapolis Motor Speedway, a center for great apes in Florida, name plates at a Holocaust museum in Illinois. The Alaska pipeline should have been up there too. On the elevated sections of the pipeline, the insulation is shielded with galvanized steel. A plaque displayed the AGA's motto: "Protecting steel for generations."

Rahrig invited me into his office and went straight into the turf battle among galvanizers, painters, and weathering-steel advocates. "The paint guys have an army!" he began. "They're loaded—they're fifty times bigger than we are. They have distributors everywhere. The biggest paint company does six, eight billion dollars a year. The largest galvanizer does three hundred and fifty million a year." Rahrig, of middle age and middle size, with a decent-sized neck, spent years working for U.S. Steel, and believes fiercely not just in steel, or American steel, but in American steel coated in zinc. "We use forty percent as much galvanized steel as Europe. They have a much higher sense of preservation than we do. They have that appreciation for making things last. In Europe, they're not enamored with color—everything is just gray."

Rahrig wore a white jacket, blue golf shirt, dark slacks. He'd have fit in on a stage in Vegas. He said, "The public doesn't know anything about galvanizing. It's an invisible product. I think they believe that rust is inevitable and unavoidable. If we had a budget of a gazillion dollars, we could convince 'em otherwise." To convince 'em, or at least some of 'em, Rahrig told me we'd shortly be visiting the offices of the engineering firm Parsons Brinckerhoff in downtown Denver. The AGA regularly presents to architectural and engineering firms, because such firms need continuing-education credits.

Today's presentation, Rahrig told me, would be by Kevin Irving of AZZ Galvanizing Services, the country's largest galvanizer. AZZ runs thirty-three galvanizing plants throughout the country, including one just north of Denver. Rahrig explained the approach that Irving would take to convince engineers about the value of galvanizing. "We start with a definition of the problem, which is rust." The extent of the problem would be conveyed via photos of rusty bridges, rusty railings, rusty posts, rusty beams. "We define the problem, give them data, let them make the decision," he said. Of Irving, Rahrig promised, "He's very enthusiastic. He's like Tony Robbins."

We got into Rahrig's car and headed north on I-25. As he drove, Rahrig pointed out galvanized structures.

"Everything along the roadside is galvanized: signs, posts, guide rails. I don't think people notice that."

"All these overpasses were galvanized and painted. They wanted that green."

"License plates: galvanized sheet metal."

We drove past chain-link fences and fence posts: galvanized. We drove past an electrical substation: galvanized. We stopped at a sandwich shop, and I pointed out the stainless steel countertops. Rahrig said, yeah, meat, fruit—anything acidic—needs stainless steel. Then he pointed at the bread racks. They were galvanized.

As we continued, Rahrig moaned about the ungalvanized things he saw. "See that railing? Rusted."

"Scaffolding—always painted yellow. It's a disposable item. It costs more to repaint it than to build it in the first place."

"The stairs in the parking garage at the airport—you seen them? There are four in the corners and one in the middle. DIA was built in 'ninety-four, so they were painted in 'ninety-four, sandblasted, and repainted in 'oh-one, and this year, they took 'em down and replaced 'em. They just looked awful. Now they're galvanized. Garages are moist, even in arid Colorado, so it's not a good environment for paint." He said that paint only lasts a year if it's always wet. Then he looked at me and asked if I had any idea how hard and expensive it is to paint and sandblast inside a parking garage at a busy airport.

"If the taxpayers knew what idiots . . ." He started over, holding back his Tea Party steam. Rahrig hates government, reserving a special loathing for DOT employees. He calls them myopic, lazy, not very smart. He says they're reluctant to do things a new way, by which he means galvanize, rather than paint, paint, paint. "The DOT, they don't care. It's tax dollars, they don't care. There's six hundred and fifty thousand deficient bridges in this country. What does that tell you? They don't have the budget to maintain 'em." What he means is: they're wasting their money on paint. Paint, he says, is government propaganda. "I'd rather have ten galvanized bridges than fifteen painted bridges."

In its battle against paint, and its armies built by paint giants PPG Industries and Valspar, the AGA has produced a fact sheet called "Hot-Dip Galvanized Steel Vs. Paint." It's arranged like a rundown of two candidates from opposing political parties. Not unintentionally, the galvanizing party is represented in green; paint, in blue. On ten issues, a voter can compare their relative positions, and pick the winner. Galvanizing, for example, doesn't require special handling, or field touch-up, or good weather to apply. Not so for paint. Galvanized coatings are thick, hard and resistant to scratching, and bonded to steel with ten times more force than paint. Galvanized coatings can handle higher temperatures and last seventy-five years; paint, only fifteen.

The kicker, though, is the comparison of costs. Over thirty years, a 250-ton bridge on the East Coast, for example, primed and painted with epoxy or waterborne acrylic or urethane paint, will need at least one touch-up and repainting. Painted with three coats of latex paint, the bridge will require more than twice as much maintenance, all of which is costly. Galvanized, the bridge will require . . . nothing. Though the initial costs are comparable, the overall costs are not. According to NACE, which published the analysis, the overall cost of building and maintaining a galvanized structure is anywhere from one-half to one-third that of building and maintaining a painted one.

Rahrig later explained his strategy. "We're increasing the size of the pie by going after concrete and trying to increase market share by going after paint," he said. He said he wished that he could create the demand for galvanizing from the bottom up, from taxpayers, but said he'd need "the advertising budget of Procter and Gamble." The AGA does not have the budget of P&G. The AGA places half-page ads in magazines such as *Architectural Record*, *Civil Engineering*, *Structural Engineering*, *Engineering News-Record*, and *Modern Steel Construction*. Joe Taxpayer does not read *Modern Steel Construction*. Rahrig, meanwhile, writes articles for *Roads & Bridges*, *Bridge Builder*, *Parking Professional*, *Parking Today*. It is hard to imagine *Parking Today* having a sizeable readership. He is a good salesman, through and through.

Though zinc was used more than a millennium ago in China, India, Europe, and ancient Greece—where it was known as "false silver"— galvanizing was not described until 1742. Paul Jacques Malouin, a French chemist, then told the Royal Academy of Sciences, "It is not as easy as one imagines." According to a contemporaneous Welsh bishop, Richard Weston of Llandaff, the process wasn't so tricky at all. Weston of Llandaff described galvanizing iron saucepans thus: "The vessels are first made very bright of salammoniac and afterward dipped into an iron pot full of melted zinc." That's pretty much still the procedure.

Of course, the process wasn't called galvanizing until 1837, when another Frenchman, Stanislaus Tranquille Modeste Sorel, took out a patent for the procedure. In his application, he gave credit where it was due, referring to "the important discovery by Galvani and Volta that electricity is generated through contact of dissimilar metals," one of which "was always preserved from oxidation." He credited Humphry Davy and his copper ship experiments, but wrote, "[T]he method that I propose . . . is quite different . . . This method consists of completely coating the surface of iron with a layer of zinc. First the iron is cleaned and after being immersed in hydrochloric acid or a solution of salammoniac, it is plunged into a bath of molten zinc. Iron prepared in the manner just described is preserved from rust."

By 1850, British galvanizers were using ten thousand tons of zinc a year. By 1870, America's first galvanizing business, the Jersey City Galvanizing Co., opened. It was founded by three pipe mill men. In just five years, the trio paid back its initial investment eighteen times over. When the Brooklyn Bridge was finished in 1883, its four main cables had fourteen thousand miles of galvanized wire, replacing oil, grease, and paint as the new standard. The first transatlantic telegraph cable was galvanized. The first barbed wire was galvanized.

By 1920, after a world war described as "20 percent fighting and 80 percent engineering," the American Zinc Institute was engaged in the publicity campaign that the AGA wages to this day. At a gathering in Chicago, the galvanizing group made its pitch with the help of a newspaperman from Missouri named P. R. Coldren. "Zinc is not nearly as widely known as it should be," Coldren announced. "It isn't a metal that gets into print easily, because there isn't any particular romance wrapped up in it. By nature it is a dull and prosaic metal. Its very color is against it. Gold glitters, silver shines, diamonds sparkle, and rubies glow, but what does zinc do? . . . Nobody ever argues there is a pot of gold at the foot of the rainbow . . . No one ever said the streets of heaven are paved with zinc. No thief ever broke into a man's house, or blew up a safe, or held up an express train to get zinc. No beautiful heroine was ever tempted by a villain with zinc jewelry. Judas probably wouldn't have betrayed Jesus for thirty pieces of zinc."

This is why Phil Rahrig was headed to Parsons Brinckerhoff.

In a sixth-floor conference room, ten engineers—all men, with just one mustache—came to learn about galvanizing. Irving, a large-bellied man with a potent Chicago accent, began with a slideshow. As promised, it included photos of a rusty railing, a rusty post, a rusty beam beneath an off-ramp. "Total deterioration," Irving said. Over another image, Irving had placed the phrase "Got Rust?" The engineers did not laugh. Irving showed a photo of pigeon poop on a beam, and another of a beam riddled with rust holes, as if hit with a shotgun. Irving said, "If you keep the barrier on steel, will it rust? No!" Then Irving said that the cost of corrosion was equivalent to building 562 Sears Towers every year.

Since much of that cost is tied to bridges, Irving cited a unique span northeast of Indianapolis, on I-69. The Castleton Bridge was built with a galvanized structure under its northbound lanes and a painted structure under its southbound lanes. Since it was completed in 1970, the southbound side has been repainted in 1984 and 2002. By 2012, Irving said, Indiana had spent more to maintain the bridge than it had to build it. Irving said he inspected the northbound half, and after forty-two years, it's zinc coating was still, on average, 6.8 mils, or thousandths of an inch, thick. "That's outrageously good," he said. The AGA expects travelers headed north to be able to use the bridge for sixty years.

The AGA is pretty sure that the first fully galvanized bridge in the country is Michigan's Stearns Bayou Bridge, built in 1966. It is four hundred feet long, spans fresh water, sees moderate traffic, and is—in the parlance—subject to salting in the winter. When the AGA last inspected the bridge, in 1997, it concluded that its beams were in "very good shape," its bolted connections "good," neither showing any signs of rust. Average coating thickness was 6.3 mils. Only the handrails showed slight staining. The AGA figured the bridge would last sixty-six years.

In the middle of the rust belt, Ohio seems to have gotten the AGA's message. It has over 1,600 galvanized bridges, more than any other state. Chicago just built 8. Renovating the Tappan Zee Bridge, New York turned to galvanizing, and saved $3 million. Pittsburgh, on the other hand, learned the hard way that concrete was no rival. Just east of downtown, a flat bridge was built beneath a concrete bridge, solely to catch the crumbling concrete from the bridge above. "Concrete comes with a couple of guarantees," Irving said, smirking, as he showed a photo of the structure. "It will crack, and it won't burn." Again, nobody laughed.

Next slide. "Here's a bridge—it's a piece of junk!" But wait, he said. There was no need to scrap that steel. "You can clean it, coat it, and put it back up." Even the old railings could be regalvanized. An engineer in the room grew curious. "How long does recoating take?" Irving said it takes less than two weeks to blast and galvanize an old bridge like that. He said plants are hot all the time, ready to go. All you gotta do, he said, is schedule it.

Among the 170 galvanizing plants in the United States, most of which are in the East, the Midwest, or Texas, none has a kettle larger than sixty feet long. What this means is that the longest beam anyone can hot-dip galvanize is a ninety-footer, and this is not prohibitive. This is accomplished by dipping in one end, lifting it out, flipping around the beam, and then dipping in the other end. Galvanizers call this progressive dipping. The dip is into molten, 840-degree zinc, four times as thick as maple syrup. Before dipping a pipe or tube or anything hollow, a galvanizer will blow holes in the object, to let air and water vent out, and molten zinc in and then out. Using a magnetic thickness gauge, an inspector can assure himself, once the beam or pipe has cooled, that the entire thing has been coated.

Coated is a funny word, because, really, the two metals are metallurgically bonded. If, using an electron microscope, you examine the thin layer of zinc on the surface of the steel beam, you'll see four distinct layers. From the steel up, they are known as gamma, delta, zeta, and eta. The first three layers—respectively, 75 percent zinc, 90 percent zinc, and 95 percent zinc—are harder than steel itself. The outermost layer, eta, is 100 percent zinc, and hence the softest layer; scratchable, but with difficulty. Irving picked up two brick-sized samples of galvanized steel and clacked them together. Then, describing the zinc-to-steel bond, he punched his right fist into his left palm. Boom!

While a galvanized steel beam cools over days, reactions slowly convert the zinc to zinc carbonate. To guys like Irving, this is primo stuff, worth waiting for. It's worth waiting for because it bonds very well with paint, and a beam that has been galvanized and painted—what's known as duplex—benefits from something of a synergistic effect. An engineer using duplex-treated steel can plan on about twice the life expectancy he'd otherwise expect, because rust doesn't form immediately below scratches in the paint. In fact, rust doesn't form when the zinc coating is scratched, because galvanized steel heals itself cathodically. It can tolerate wounds up to a quarter inch. "Zinc is the best primer in the world," Irving said. The main cables on the new San Francisco-Oakland Bay Bridge are duplexed.

Ending his pitch, Irving answered a handful of questions. He said that the cost of galvanized rebar was the same as that of epoxy-coated rebar and said that six states, including Florida, Virginia, and Oregon, no longer allow epoxy-coated rebar in highways because of cracked coatings leading to rust. He described how to weld and repair galvanized steel, and didn't seem to frighten any of the engineers. He said that the auto industry has started to use galvanized sheet metal. "Remember when you used to get your car Rusty Jonesed? I had two cars that I got Rusty Jonesed! Now, with galvanizing, Ziebart is gone!" Actually, Ziebart still applies aftermarket rustproof undercoatings to cars, while its former competitor, Rusty Jones, went bankrupt in 1988.

Then Irving compared California to Italy. "They're the same size. California has seven galvanizing plants. Italy has a hundred and thirteen. You go into grade schools, and the kids know about galvanizing. Europe galvanizes fifty percent of their steel. We do, what, six? We're a disposable society." He sounded like Dan Dunmire. He said, "If Roy Rogers was alive today, he'd want his horse Trigger in a galvanized pen." This was strange, as nine out of the ten engineers in the room were too young to get the reference. I thought he should have repeated something he said earlier: "Our kettles are always on."

8

TEN THOUSAND MUSTACHIOED MEN

In 1997 a corrosion engineer named Rusty Strong had a rust problem. He was on his way back to Houston from a corrosion conference near Chicago, and after deplaning, he took a shuttle to the airport parking lot where his black Nissan truck was parked. Before deshuttling, he could tell that something was wrong. His truck was crushed—the cab banged in, the windshield shattered. Incredulous, he asked the shuttle bus driver what had happened. The driver refused to make eye contact. "I think a pole fell on it," she mumbled nervously. "It was an act of God." Rusty was steaming, particularly since nobody had bothered to cover the hole in the cab, and after a few days of rain, the floors were soaked. He took the shuttle back to the parking lot toll gate and called a tow truck.

Late the next morning, in his wife's car, he swung by his office, grabbed a camera and micrometer, and returned to the lot. He began investigating. The twenty-foot light pole that had fallen on his truck had been removed, but the base of the pole, four inches in diameter, was easily visible on a concrete pedestal one foot off the ground. The base of the pole was heavily rusted on the inside, because the weep hole that was supposed to let water out had been grouted over. Rusty took photos and measurements. Then he began inspecting other poles in the parking lot, taking more photos and more measurements. That's when the shuttle bus pulled up. Out came the parking lot manager, telling Rusty he wasn't authorized to take photos. They argued, while Rusty finished his study. Then Rusty asked to talk to the owner of the lot. The owner, in Florida, told him by phone that the pole had been knocked over by a tornado in a rainstorm. Rusty—informing the owner that he was a corrosion engineer who studied rust professionally—told the man otherwise. "This was not an act of God," he said. "It was a failure of man." He went on, informing the lot owner that, were the matter to end up in court, it was precisely someone like Rusty that the owner would want on his side. Rusty figured that the owner didn't buy it—a rust professional? Who'd ever heard of such a thing? After that phone call, Rusty drove home and made another phone call, to his insurance company. He told his agent that the damage to his truck was the result of a maintenance failure. To that agent he faxed an article from the journal *Corrosion* on the same rusting-light-pole phenomenon in Galveston, Texas, along with his photos. Fifteen minutes later, an insurance adjustor called Rusty. He was laughing. "This'll be so easy," he said. Now Rusty's insurance record says DO NOT CANCEL. He doesn't park at that lot anymore.

Rusty tells this story to colleagues and fellow corrosion engineers at conferences of the National Association of Corrosion Engineers, and it always elicits a similar response. That's hilarious, they say. And: I can't believe they tried to pull a fast one on a corrosion engineer.

Of the fifteen thousand corrosion engineers in the United States, most don't deal with statues, cans, air force jets, or navy ships. According to NACE (which has since rebranded its name to NACE International, the Corrosion Society), a quarter of its members work in pipeline integrity. Another 10 percent work for natural gas utilities, an additional 9 percent work in oil and gas extraction, and still more work for refineries—which means that about half of all corrosion engineers work in oil and gas. Corrosion engineers who don't work in oil and gas most likely work in some transportation-related field, on planes (and spacecraft), or ships, or cars, or roads, or bridges, or docks. Some work in mining, or paper processing, or manufacturing. Many work for water, electric, or sewage utilities. NACE International counts among its corporate members the City of Los

Angeles Department of Water and Power, the Baltimore Gas and Electric Company, Colorado Springs Utilities, Knoxville Utilities Board, Santa Clara Valley Water District, West Virginia Department of Transportation, Pacific Gas and Electric Company, and the US Bureau of Reclamation.

Many make chemicals, or metals that resist high temperatures, or biomedical implants. Metallic implants are made mostly of biocompatible materials such as stainless steel, platinum, or titanium, though the brother of corrosion consultant extraordinaire Bob Baboian had a plate of tantalum put in his head after suffering a wound in World War II. Nonbiostable implants corrode and present as arthritis. The latest stents, used to keep narrow arteries open, are made of nickel alloys, platinum chromium alloys, and cobalt chromium alloys, and some made of niobium are in development.

Many corrosion engineers, conducting their research at educational institutions, also teach. The majority work for almost 1,500 different companies, including 3M, BASF, Dow Chemical, General Electric, Halliburton, Honeywell, Hyundai, Northrop Grumman, and Siemens, as well as Corotech, Cortec, Cortest, Corr Instruments, CorroMetrics, Corrpro, Cor-Pro, Corrodys, and Corrosus, which sounds like it ought to team up with T-Rex Services. More than a handful work at government labs, including Los Alamos and Sandia National Labs, or the Naval Research Laboratory, or at the Nuclear Regulatory Commission, or at NASA. Some work at private corrosion labs, solving problems for organizations lacking their own corrosion engineers. These private labs, by the way, from Nevada to Delaware, were generally not amenable to showing around a guy writing about the industry.

A handful of corrosion engineers work as father-and-son teams. At Corrosion 2012, NACE International's sixty-seventh annual conference and expo, which was attended by six thousand members of the corrosion business, I met a young corrosion engineer named Ryan Tinnea. Ryan Tinnea is the son of Jack Tinnea, who is also a corrosion engineer. Jack has a mustache. Ryan does not.

On the expansive floor of Salt Lake City's convention center, I followed the younger Tinnea to the booth of a vendor selling plastic rebar. We walked past rows and rows of booths, from ten-by-ten-foot stalls to twenty-by-thirty-foot islands, selling $25,000 handheld X-ray fluorescence detectors,* paints that change color with heat, and corrosion inhibitors so potent that one drop keeps steel wool in a jar of water unblemished. As we walked, Tinnea told me that he thought the fraction of corrosion engineers working in oil and gas was more like two-thirds. Surrounded as we were by booths tailored to the wants of big spenders, this impression was understandable. At the booth of the rebar vendor, Tinnea asked the salesmen about the mechanical characteristics of their stuff. Pseudoductile, they called it. It was not good enough for Tinnea. If not ductile, it was not okay in earthquakes. A fault runs directly through the middle of the city where he and his father work, about a mile south of the office of Tinnea & Associates. He and Tinnea the elder work in Seattle, on everything from the aquarium, to the opera house, to Piers 58, 59, and 60.

About 8 percent of corrosion engineers fly solo as corrosion consultants, providing independent corrosion expertise, often in service of litigation after some calamity: delaminating space shuttles, leaking pipelines, out-of-commission offshore oil rigs, or houses built with tainted Chinese drywall. They are not short of work. I found Tinnea the elder in his office on a Saturday morning. On a survey sent by NACE International to American corrosion engineers, one commented: "too much work to do with not enough time." John Scully, the editor of the journal *Corrosion* (which is published by NACE International), put it thus: "Some people worry about job security. Corrosion engineers will always have job security." Tom Watson, who served as NACE's president from 1964 to 1965, put it even better. In June 1974, he wrote this poem, titled "Rust's a Must":

> *Mighty ships upon the ocean*
> *Suffer from severe corrosion;*
> *Even those that stay at dockside*
> *Are rapidly becoming oxide.*

---

*I checked out the XRF detector on a book of metal samples. It works because every element, aside from the noble gases, from magnesium on up fluoresces uniquely when bombarded with X-rays. According to guys on the floor of the convention center, the drill-sized tool is "the shit" for identifying the composition of, say, unknown pipelines. The tool has one detector but sends three or four beams to be sure it's accurate. It works in seconds. It's way easier than cutting out a sample and sending it to a lab. Though I need the tool in my daily life about as much as I need a bulldozer, I want one.

Alas, that piling in the sea
is mostly Fe_2O_3.
And when the ocean meets the shore
You'll find there's Fe_3O_4.
'Cause when the wind is salt and gusty
Things are getting awful rusty.
We can measure, we can test it,
We can halt it or arrest it.
We can gather it and weigh it,
We can coat it, we can spray it.
We examine and dissect it,
We cathodically protect it.
We can pick it up and drop it,
But heaven knows we'll never stop it!
So here's to rust, no doubt about it,
Most of us would starve without it.

Watson, as far as I can tell, was the funniest corrosion engineer in history. At a conference in Toronto during his tenure, he accidentally lit a block of magnesium on fire, and it burned a hole through the floor of a Holiday Inn. Toronto, apparently, did not hold it against the group. Because corrosion engineers aren't unlike other engineers—as the joke goes, even the most extroverted one looks at your shoes while you talk to him—Watson is a great exception.

The other great exception is O. Doug Dawson, of AutoChem. In 1972 he presented a paper at the Australian Corrosion Association's thirteenth conference called "Sex and Corrosion." Half seriously, he alleged that the mechanisms of aqueous corrosion were analogous to those of sexual reproduction. Courtship, love, conception or contraception, pregnancy or abortion, and gestation all fit into his schema. The way he saw it, the patterns observed between the range of metals on the galvanic series (between magnesium and gold) seemed comparable to the patterns observed by scantily clad beachgoers (between surfer dudes and bikini babes). In the world and on the beach, factors at play in coupling included propinquity, exposure, and "local protrusions." Bulges and curves, he wrote above the line drawing of a female profile, should be avoided.

There's always a reason
A time or a season
Conductive to copulation.
So if you believe
Metals really conceive,
You ought to expect some gestation.

The 1970s were a different time.

Most corrosion engineers today are serious, conservative, and not especially social or rowdy or funny. Inquiring about rust jokes at Corrosion 2012, I got blank looks. "I can't think of any rust jokes," one corrosion consultant, known for his gregariousness, told me later. "Just wife jokes." In two years of asking for rust jokes, I never found one. The handful of times rust has appeared in Sunday comics, the punch line was either from an auto mechanic saying, "Hey, you might wanna deal with this" when it was way too late, or a husband telling his wife, "I used your face cream to take the rust off my lawnmower—it works great!" The wife doesn't look so contented.

I overheard a brief conversation in which one corrosion engineer told another that, the night before, a bunch went to a karaoke bar and started singing.

"Hey, who are you guys?" asked someone at the bar.

"We're corrosion engineers," he responded.

Then the bar cleared out.

Some 93 percent of corrosion engineers are men. There's no official data on what fraction of them have mustaches, but my estimate is high. The field is full of stable old-timers: 40 percent have been working in corrosion for twenty or more years, most of those at the same big (five hundred or more employees) company. A clue to their habits may be gleaned from the degree to which NACE International's annual golf tournament exceeds the annual NACE RACE 5K in popularity. Another clue can be gleaned from Marco De Marco's winning times, which in 2011 and 2012 were not under twenty minutes.

Surprisingly, they are not especially educated: fewer than one in three have a bachelor's degree. One in ten have a master's degree; one in sixteen have a PhD. The Accreditation Board for Engineering and Technology

(ABET) does not recognize corrosion engineers as professional engineers, mechanical, civil, electrical, or otherwise. California briefly licensed professional corrosion engineers but ended the practice in 1999. NACE does not keep track of how many of its corrosion engineers are licensed professional engineers but thinks it's most of them. This seems unlikely. One in four corrosion engineers holds other professional certification, from the American Petroleum Institute, or the American Welding Society, or ICORR, the Institute of Corrosion.

Education notwithstanding, the average annual salary of a corrosion engineer is just shy of $100,000. This is significantly better than the averages for architects and engineers, as the Department of Labor sees it. Roughly 11 percent of corrosion engineers make more than $150K; 4 percent make more than $200K. Salaries are going down in Europe but are on the rise in the United States. Those who make the most work for tiny companies or huge ones, or live in Alaska, where oil flows like the Yukon.

A quarter of American corrosion engineers live in Texas, though corrosion engineers reside in all fifty states and DC. Just as many hail from the rest of the world, in 110 different countries. There's one corrosion engineer in Botswana, one in Côte d'Ivoire, one in Equatorial Guinea, one in Zambia, one in Uzbekistan, one in Macau, and one in Tahiti. They gather locally in 120 "sections"; as expected, the Houston section is the largest. Notwithstanding the Texas oil and gas clique, corrosion engineers are dispersed widely among us.

Wherever they are, they tend to wish their work—and trade—was more widely appreciated. In that same NACE survey, corrosion engineers submitted these comments:

> "There seems to be a perception that my job is just to make other
>   people's jobs more difficult."
> "Generally, the uninformed hold positions of power and routinely make
>   bad decisions."
> "It just seems in our industry we are always reacting to unprotected
>   situations instead of being out front managing the systems at a
>   protected level."
> "In a lot of circles, we are a nuisance rather than an integral part of the
>   process."

"People often cut the budget and then wonder why they have an issue five years later."

Kevin Garrity, NACE International's president from 2012 to 2013, told me he knows three people who left their jobs because their bosses ignored their work. Ray Taylor, the head of the National Corrosion Center who said corrosion was "sort of the wart on the ass of the pig," elaborated. "This is not a sexy thing to get involved with. They put it off. They say, 'We'll just wait a little bit.' It just goes on and on and on. It's been surpassed in the science arena by other areas, but we've forgotten—we haven't even finished the basics yet. So many things have been put off to the side, and we haven't gotten to them. Okay, we've got all these people who do accounting and so on, but let's do a life-cycle analysis. If we just let this go, and break, and repair it then, is that cheaper than if you did a little corrosion control? People haven't done a good job with that."

Though the United States is home to dozens of men named Will Rust—writers and lawyers and graphic designers and programmers and sales managers among them—not one is a corrosion engineer. George Washington Rust, who came to my town 140 years ago to recover from tuberculosis, knew finance and cattle but not corrosion. Neither Russel Bits nor Russell Parts is a corrosion engineer, and Russel Auto Parts, for that matter, seems like a bad name for a business. Rusty Auto Parts? Russell Stough does not pronounce his name Rusty Stuff. The most popular first names of corrosion engineers are John, David, Michael, Robert, James, William, Richard, Mark, Paul—as if their calling is biblical. More corrosion engineers are named David Miller than anything else. Michael Jones, John Wilson, and Richard Smith are the next most common. The best clue to their diversity comes from their last names, the most popular of which are Smith, Wang, Zhang, Johnson, Li, Lee, Kim, Williams, and (fittingly) Brown. Three corrosion engineers are named Mohammed (or Mohammad) Ali. One corrosion engineer in Ohio, named Steven, who has a PhD in mathematical statistics, really is Dr. Rust. Another, in Florida, who has a PhD in metallurgical engineering, and the surname of Heidersbach, calls himself Dr. Rust. Ten corrosion engineers are named Rusty. Rusty Strong, of noncancelable auto insurance, believes he has the best name in the industry. "You know *Catch-22*," he said. "I'm like Major Major Major." Rusty made a conscious decision to put his nickname on his business card when he joined NACE. It got him onto the board.

NACE's roots lie so deep within the oil and gas industry that the organization has struggled to drop the reputation ever since. Eleven oil and gas men founded it in 1943 to study the determent of corrosion in pipelines. The mustache-free founders elected as their first president R. A. Brannon, of Humble Pipe Line. So as not to appear singularly devoted to the oil and gas industry—an entirely warranted assessment—the board selected as chairman of its technical committees a metallurgist named R. B. Mears. He had a PhD from Cambridge University and was chief of Alcoa's chemical metallurgy division. His neatly parted silver hair, rimless glasses, and effeminate lips gave him a scholarly demeanor; his colleagues, on the other hand, appeared thuggish, as if planning to give corrosion a beating. Mears looked more like a preacher than an oil and gas man, and this suited NACE well. Yet well into the 1960s, NACE's bent as an oil-and-gas affair persisted. Dale Miller, who began working as an editorial assistant at *Corrosion* in 1958, told Lyle Perrigo, the author of a slim history of NACE, that he didn't realize NACE wasn't solely pipeline oriented until 1966, by which point he'd been working there for eight years.

Though NACE grew rapidly—gaining more than a thousand members in its first five years, and another three thousand in the subsequent five—the organization had difficulty recruiting. Frans Vander Henst, secretary of NACE's technical committees from 1958 until 1965, told Perrigo about the military's reaction: "[I]t was a surprise that there was actually something you can do about their problems." He recalled the solution devised by an officer in Guam to keep jeeps and planes from rusting away. "He did not know what to do except to run them off the runway and drop them into the ocean." Henst collected addresses of military bases, and to them, attention maintenance officers, sent information. They came to meetings, slowly. Plumbers, though, were a different breed. "We have not made any inroads with the plumbers yet," Henst recalled. "They kept telling us they did not want to solve problems because about fifty to sixty percent of their work was in repairing systems. They were very, very adamant that they did not want to solve the problems." NACE does not count among its corporate members the American Society of Plumbing Engineers, the International Association of Plumbing and Mechanical Officials, the Plumbing-Heating-Cooling Contractors Association, or the United Association of Journeymen and Apprentices of the Plumbing and Pipe Fitting Industry of the United States, Canada and Australia. At the annual conference, I did not see a single plumber.

NACE, a $25 million nonprofit, makes money a few different ways. One-eighth of the dollars it collects come from individual and corporate members, including DuPont, Bechtel, Sherwin-Williams, BP, Chevron, ConocoPhillips, ExxonMobil, and Shell; Abtrex, Aztech, Exova, Lintec, Sprayroq, and Termarust; Dan Dunmire's office in the DOD, and Tinnea & Associates. Another sixth it collects when these companies pay $25 per square foot for booths at annual corrosion conferences, selling their rebar, paints, and badass X-ray fluorescence detectors. NACE makes half of its revenue through its corrosion courses. The photo on the cover of its 2012 course catalog is one of my favorites in the industry. It shows nine guys sitting and standing around a table, studying for a test. All five on one side of the table have mustaches, as if in battle with the bare upper-lipped four on the other. NACE's courses fall more or less into three categories: basics, coatings application or inspection, and cathodic protection. Some are pipeline themed or marine themed or wastewater themed or nuclear themed. The generic five-day courses cost about a thousand bucks; the industry-themed ones cost nearly double that. Daily, that's more expensive than an Ivy League education, which comes with room and board.

In 2010 NACE reported that ten of its staff earned more than $100K, some nearly $200K. The organization has also officially begun lobbying. In Pennsylvania and Ohio, it supported the campaigns of two Republican representatives running for office. Rust could be a progressive issue, but interwoven as it is with Big Industry, it leans right.

NACE also sells a wide variety of industry-specific textbooks. If you're looking for a rust joke, these publications are a certain dead end. Most of the books cover the particulars of corrosion science, the behaviors of various metals, the oil and gas industry, or paints. Titles include *The Water Dictionary, Deep Anode Systems, Practical Chlorine Dioxide, Concrete: Building Pathology,* and *Coatings Tribology.* The best coatings book is hands-down *Fitz's Atlas of Coating Defects,* an expensive but easy-to-use visual guide to the various ailments that may afflict a coating. A coating may be cheesy, checked, rippled, wrinkled, peppery, seedy, saponified, crocodiled (or alligatored), cratered, crazed, cobwebbed, crow's-footed, or cracked, like mud or stars. It may look, technically, like an orange peel. It may appear flocculated, or have fish eyes, or be flaking. It may be blistering, bubbling, cissing, disbonded, delaminated, pinholed, peeling, or just plain undercured. Given that level of precision, it's not surprising that *Corrosion Testing Made Easy* is a six-book set costing $500. Most of the textbooks that NACE sells cost around $100, and a few approach $1,000. A NACE employee told me that they are very profitable. The organization's various publications contribute the final sixth of its revenue.

One of the cheapest books that NACE sells is a cartoon book called *Inspector Protector and the Colossal Corrosion Fighters*. It was published in 2004 as an educational booklet for kids in the vein of Marvel Comics. The good guys total five. There's Inspector Protector, masked and caped in blue, with Superman-like abs. There's Dr. Forbidden, bearded and bespectacled, in a white-lab coat, and in possession of a jar of steaming green corrosion inhibitor. There's SuperCoat, the brunette "coatings gal," with two tanks strapped to her back and a very assertive paint gun. There's Captain Cathode, a sort of Terminator-like android, and his smaller, stouter Awesome Anodes. Finally, there's Smart Pig, who looks like a cross between a college football coach and an actual pig, with a radio collar and a red light on his head.

Together, the good guys fight Count Corrosion, who has a sickly green pallor, greasy black hair, and an unforgiving, Transylvanian facial structure stuck in a permanent grimace. Count Corrosion works alongside his evil swarm of Grubbz—giant six-legged termites who crouch like Sméagol in *Lord of the Rings* and go crunch! and chomp! as they gobble away.

In sixteen pages, the protagonists go from the Statue of Liberty to the Golden Gate Bridge to Edmonton, Alberta, to—cliffhanger, here—an airplane, about to lose a wing.

Sploosh! Splat! Whoops! Thwack!

Biceps bulge. Clenched fists fly. Weapons are raised.

Our hero corrosion fighters appear young, strong, and vibrant. They're fast acting. They live not in Houston but in some sort of flying fortress. They report to no bosses, obey no budgets, put up with no regulations or layers of company or government bureaucracy. Our heroes don't sport compensatory mustaches or cell phone holsters or pocket protectors. They wear their underwear on the outside. Nor do they attend dull meetings in Salt Lake City. The book ought to have been the best-selling product at the convention.

9

PIGGING THE PIPE

MILE 0

Three hundred miles into the Arctic, at the northern terminus of the Trans-Alaska Pipeline System (TAPS), a forty-one-year-old engineer named Bhaskar Neogi hummed Beethoven. He was sitting in the Maintenance Tech Office of Pump Station 1, thinking about rust. Oil and gas men like Neogi tend not to like the word *rust*, though. They call it black powder, and they call corrosion engineering "integrity management." Officially, Neogi is the pipeline's integrity manager. He is responsible for keeping the pipeline intact, whole. Most pipeline operators employ integrity managers, but most pipelines are not like the Trans-Alaska Pipeline System. From Prudhoe Bay to Prince William Sound, TAPS stretches eight hundred miles, which leaves Neogi accountable for one of the heaviest metal *things* in the Western Hemisphere, through which the vast majority of Alaska's economy flows. Daily, the four-foot steel tube spits out $50 million of oil. Even for an engineer as bright and focused as Neogi, such accountability is awesome. That's why, in March 2013, he was humming Beethoven. He was anxious. After more than a year of preparation, he was about to launch a $2 million rust-detecting robot through the pipeline, and he was worried about the robot's ability to perform, let alone survive, the long journey.

The robot was a "smart pig," sixteen feet long and more than ten thousand pounds, and suggestive of a giant centipede. In the cavernous manifold room on the other side of Pump Station 1, it sat on a tray just inside an orange bay door. Outside the door, it was ten below and windy. Inside, where it was warm, four technicians from Baker Hughes, the pig's manufacturer, wrapped up a third day of checking and double checking and triple checking its componentry. Among other things, in the front segment of the pig, between two yellow urethane cups, they checked 112 magnetic sensors mounted in between 112 pairs of magnetized brushes. These sensors would detect the magnetic field induced in the pipe as the pig, propelled by the flow of oil, traveled through it. Given any kind of anomaly in the half-inch steel—a pit, a ding, a thin spot—the field would change, and the sensors would capture this and record it on a hard drive. Inch by inch, the sensors would capture this information; Neogi hoped they would capture all seven billion square inches of the pipe. That's 1,200 acres. Using all that data, Neogi would determine the most vulnerable spots on the pipeline, dig them up, and repair them before they became leaks.

Neogi was humming because no matter how extensively the technicians double checked, even the most advanced pig can't perform its inspection if the wall of the pipe is covered in wax. Wax, a natural component of crude oil, keeps the magnetic brushes and sensors off the steel wall. The consistency of lip balm or mousse, it plugs up caliper arms that measure the shape of the pipe, and snags odometer wheels. Wax renders smart pigs senseless, leaving them blind, dumb, and amnesiac. Nor can a pig survive a violent voyage. Too fast, and sensor heads melt or crack. Too rough, and the magnetizing brushes wear down. Too jarring, and the universal joint between the pig's two segments comes apart, wires snap, and power to the magnetic flux sensors is cut off. Poof goes the data, months of work, and millions of dollars—leaving engineers with a pipeline in indeterminate condition, regulators unhappy, and the public at risk. Wax accumulates when the oil cools below 75 degrees, and long, slack sections, where the pig can barrel down mountain passes at high speed, manifest themselves when there's not much oil flowing through the pipe. Neogi was well aware that it was winter, and that the flow of oil through TAPS was as low as it had been. It was not the best of times to pig. Yet he had faith.

"I want to know exactly what's going on," Neogi said. He had dark eyes,

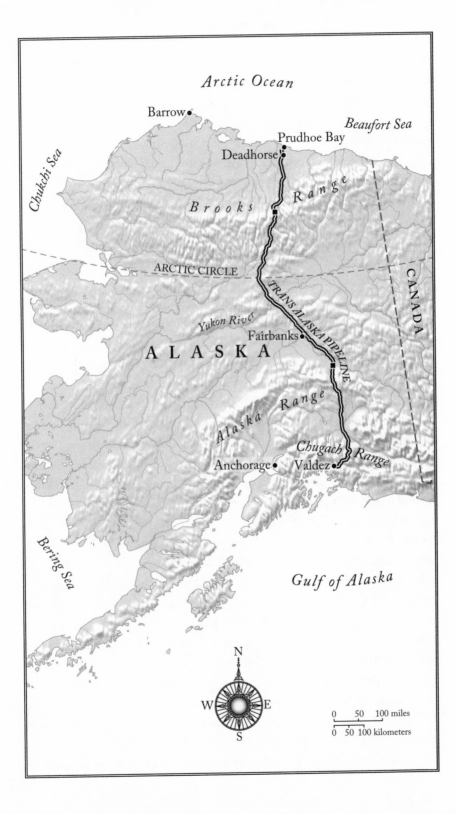

short, dark hair with flecks of gray on the tips, and was wearing a navy blue fire-retardant shirt. Built like a soccer player, he spoke in a quick staccato, with a voice that was reedy and delicate, and with a noticeable Indian accent. "The great fear is that there's some corrosion that we can't see," he said. "If you can't see it, you can't do anything about it because you don't know it's there. Whatever we can assess with our pigging, we can fix." His manner was formal but blunt, almost brusque. From the pig, he said, he yearned to get what he called the pulse of the pipeline. "You break your tool, and you don't get data," he said. "We've done so much preparation—everything we can—but it's possible we'd have to redo the run." This wasn't idle conjecture. On account of wax and low flow rates, in the last dozen years, half the smart pig runs have failed.

That afternoon, shortly before Neogi began humming, the technicians spent forty-five minutes testing the pig's lithium batteries, each of which weighed twenty-five pounds, cost $3,000, and was not rechargeable. They checked the hard drive in the rear of the pig, and they repositioned one of the pig's two transmitters, which would make tracking it during its eighteen-day journey possible. Meanwhile, Neogi and several executives from the pipeline's operator, the Alyeska Pipeline Service Company, ran through the second of three final "go/no-go" meetings. Though pigging the pipe is a routine maintenance procedure, it is by no means casual: hazards abound in opening up the end of a vessel full of pressurized crude, putting a solid object in a pipeline designed for the transport of liquid, and trying to follow said object through the Arctic in the winter. Plenty could go wrong. In one of the more recent pigging mishaps, known as the gas excursion incident, Pump Station 1 nearly blew up. That was BP's fault, but Alyeska took heed. More recently, a pig was sucked into a relief line at a pump station midway down the line. That the relief line was only sixteen inches in diameter, and guarded with pig bars, was not a sufficient deterrent to the forty-eight-inch pig. This has happened at least a half dozen times. When it happened in 1986, and the pipeline was shut down while the pig was extracted, that meant more than a quarter of the nation's oil wasn't moving toward California. Pigs have made it all the way to Valdez, Alaska, only to be ingested in relief lines there. Other pigs have damaged the pipeline, or gotten stuck in it and been destroyed during their extraction.

Discussing conditions, logistical readiness, and safety, a handful of directors and vice presidents now considered whether or not to run this pig. Over the phone, Neogi reassured them that everything was on track. Hydraulic procedures for getting the pig smoothly down the slack mountain passes were ready. All other pipeline maintenance had been halted. Up and down the line, all eyes were on the pig. It still had a green light.

That evening, after one final check, the Baker Hughes technicians turned on the pig and then, using a pair of overhead cranes, loaded it into the launch tray at the end of the pipeline. A half dozen pump station technicians assisted, as did Neogi. They wore Tyvek protective suits and gas masks and worked slowly and carefully. They called the pig "the tool." When they opened the end of the pipeline, vapors seeped out. Gas detectors blinked, and explosion alarms went off. The ventilation system whirred. A stubborn beast, the five-ton tool did not make things easy. Just sitting in the launch tray, it required hydraulic jacks to support it. Shortly before midnight, the men finished loading it, and then closed and secured the end of the pipeline shut. Then they waited. They planned to launch the tool at seven in the morning, exactly twelve hours behind a red urethane pig of lesser intelligence. That pig, like a giant squeegee, was scraping the line clean. It was the last of nine such scraper pigs that, by Neogi's design, had been shoved down the pipeline in the previous six weeks. Neogi had kept track of how much wax these pigs had pushed out in Valdez, and graphed it. From 1,200 pounds, the mass had dropped to 400. The line was as clean as it was going to get, primed for inspection. It was ready for the smart pig.

Neogi woke at five o'clock. On the drive from Deadhorse to Pump Station 1, the flares of distant oil fields were barely visible in the dark. He drove through two security gates. Next, he plugged in his truck beneath the ice-encrusted radio tower, walked through the refrigerator door, and took off his fur-lined parka. Then he called Alyeska's main office, in Anchorage, to participate in the final go/no-go meeting. The men talked slowly. A storm seemed to have passed; the windchill was no longer 50 below. Their paperwork was in order. No new safety concerns had emerged. For twenty minutes they deliberated, and Neogi was never surprised—a good sign. The pig launch was on. Nobody noticed that it was the Ides of March.

A few minutes before seven o'clock, three station technicians opened valves in the launcher and diverted oil behind the pig. It didn't budge. The techs, all mustachioed, sent more oil behind it. Still it wouldn't budge. The pig—the biggest and most capable that Baker Hughes makes—was by far the heaviest that had ever been put in the Alaska pipeline, and the entirety of the pipeline's flow pushing against it didn't compel it to move. In the control room, where Neogi watched a little red light on a panel that would indicate the pig's departure, he slouched, biting a fingernail on his left hand. His shirt was untucked. He looked like a family member in a hospital waiting room, edgy with anticipation. That this was the first pig under his management also contributed to his anxiety. Over the radio, he listened as the techs requisitioned more oil from two giant tanks. Up the flow went, to 600,000, 700,000, 800,000 barrels per day, and still the pig didn't budge. It didn't budge until propelled by 865 pounds per square inch of crude, flowing at a rate of 840,000 barrels per day—more than half again the typical flow of oil in the pipeline. It finally took off at a quarter after seven, bound to see what rust had done to the biggest, baddest oil pipeline in the world. As the pig headed south into the barren, wintry immensity of Alaska's North Slope, it sounded like a train. All Neogi said was "All right."

For two decades, the Prudhoe Bay oil fields—Sadlerochit, Northstar, Kuparuk, Endicott, Lisburne—have been declining steadily. Yearly, immutably, they produce 5 percent less oil. The result is that TAPS now carries one quarter of the oil it was designed to carry. It comes out of the ground colder than ever and flows more slowly toward Valdez. Crude used to make it to Valdez in four days, as if running seven-minute miles. Now it walks. En route, it cools off even more and, as it does so, deposits more wax on the pipeline. A doctor would call the pipeline arteriosclerotic. While a pipeline waxes, its diameter wanes. Declining throughput makes things difficult for Neogi, but it makes them even more difficult for agencies estimating the pipeline's lifespan. The pipeline was designed to survive as long as the oil fields. Lest it clog, it must stay warm, which means that it must remain full of flowing oil. In a perverse symbiosis, the pipeline needs the oil as much as the oil needs the pipeline. As a result, while the consortium of agencies that oversees the pipeline has written that it "can be sustained for an unlimited duration," Alyeska figures that it'll survive until 2043, and the state of Alaska figures that it'll expire a bit sooner. Private consultants, hired to estimate the life of TAPS, mention only "the future" and write of "diligent upkeep" in passive sentences. The estimates all couch what no-body wants to say: the pipeline, once the largest privately funded project in America, and one of its greatest engineering achievements, is now an elderly patient in intensive care.

The companies that built the pipeline foresaw such a future and tried to avoid it. In the immediate aftermath of their 1968 oil discovery, they considered every alternative to a pipeline. They considered extending the Alaska Railroad to the North Slope, until they realized that it'd take sixty-three trains, each one hundred cars long, every day, to ship their oil. They considered trucks, calculating that they'd need nearly the entire American fleet in addition to an eight-lane highway. They looked into jumbo jets supplied by Boeing and Lockheed, turning away when it became apparent that their air traffic would exceed the combined air traffic of all the freight in the rest of the country by more than an order of magnitude. They looked into blimps. They commissioned the world's largest icebreaking cargo ship, and after it got stuck in the Northwest Passage, they seriously considered using a fleet of nuclear submarines to ship the oil, under Arctic ice, to a port in Greenland. Reluctantly, out of alternatives, they settled on a pipe-line. A steel tube winding across Alaska was ten times the rust risk of a giant copper lady standing in New York Harbor, and they knew it.

On most other pipelines, "events" or "incidents" or "product releases"—what the rest of us call leaks or spills—are most often caused by third-party damage. By this, the industry means accidents. Heavy equipment is usually to blame; pipeline ruptures are most often caused by collisions with bulldozers and backhoes.* On TAPS, since there's so little construction across the vastness of Alaska, the risk of accidental third-party damage is low. Natural hazards, on the other hand, present threats in abundance. Earthquakes, avalanches, floods, and ice floes all threaten TAPS. But what really keeps Alyeskans up is corrosion. It's the number one threat to the integrity of the Trans-Alaska Pipeline, enough to make engineers in the last frontier dream of Bakersfield.

---

*Pipeline ruptures around the country have also been caused by collisions with plows, barges, drag-racing cars, a runaway horse, a tugboat, a cement truck, and an air force jet. Ruptures have been caused by welders, construction workers, and by crews laying a new pipeline.

On account of that threat, the pipeline was outfitted with the greatest corrosion-protection features of the era. Its principal protection was its coating: paint. As a backup, a zinc strap the size of a wrist (a giant anode) was buried under the pipe. Though TAPS was, boldly, called rustproof, the defense proved insufficient. Like all coatings, the one on TAPS proved vulnerable—but Alyeska didn't learn quite *how* vulnerable for a dozen years. When it did, the company beefed up the pipeline's corrosion protection with ten thousand twenty-five-pound bags of buried magnesium anodes (recall that magnesium is the least noble metal, most willing to sacrifice itself), and a cathodic protection system consisting of a hundred-odd rectifiers spitting a low voltage into the pipe. Unlike the buried zinc strap (the condition of which and its very existence are mysterious), the mag bags and the cathodic protection system are testable: corrosion engineers can disconnect them and use probes that measure the changes in the voltage in the soil. Because rocks resist current, the cathodic protection system doesn't work well in rocky areas, leaving corrosion engineers to their final tool: coupons. On the pipeline, a coupon is a one-inch square of steel, connected to it and buried along it, serving as a surrogate. Alyeska has about eight hundred of them. But coupons don't prevent corrosion; they just help engineers monitor it. In a way, monitoring is Alyeska's second line of defense, and Alyeska does a lot of it.

Like all major pipelines, TAPS is monitored by leak-detection software, which compares the flow of oil going into the pipeline with the flow coming out the other end, and also scans for sudden pressure drops. But unlike other pipelines, it is also monitored regularly by pilots using infrared cameras to hunt for signals that the hot oil has escaped into the cold Alaskan earth, as well as by "line walkers" who hunt for dark puddles and squishy tundra along the pipeline, and by controllers watching an array of hydrocarbon-detecting and liquid-detecting and noise-detecting sensors shoved into the ground alongside it. And then there are the dozen state and federal agencies looking over the shoulders of the thousand people operating the pipeline, making it the most regulated pipeline in the world. But because a smart pig is the only way for Alyeska to determine if its pipeline is about to spring a leak before it has actually done so, and because Alyeska operates under more regulatory scrutiny than any other operator, it sends smart pigs down the line nearly twice as often as any other pipeline operator. It employs a smart pig once every three years, and has been doing so since long before federal pipeline laws stipulated it.

Thanks largely to smart pigs, TAPS hasn't suffered a corrosion-induced leak since it began operating in 1977.* Over its first thirty years, Alyeska reviewed nearly 350 potential threats to the pipeline, including dents, wrinkle bends, weld misalignments, ovalities, gouges, and corrosion pits. The majority of these problems were found with smart pigs. But pigging hasn't been a cakewalk for Alyeska. Early smart pigs weren't so bright or amenable, and since 1998, smart pigs have been stymied by wax. In fact, wax precluded Alyeska's preferred smart pigging technology: ultrasonics. Such pigs gather data by bouncing sound off the walls of the pipe and listening for echoes. The ultrasonic method, a direct measurement, is superior to magnetic flux leakage (MFL), which is indirect and inferred. After 2001, ultrasonic pigs couldn't gather enough data. The wax blocked the sound. The effort was as futile as trying to get a fetal ultrasound from a pregnant woman wearing a sweater.

Consequently, keeping the pipe clean has become a priority nearly as great as keeping it whole, because the latter depends on the former. To keep it clean, Alyeska sends cleaning pigs south weekly. The company keeps a fleet of a dozen such pigs at a maintenance yard in Valdez, and in a perpetual relay, these pigs go back and forth: up the haul road, down the line. The managerial pigs—the smart ones—wait patiently while these janitorial pigs stay busy. Before the last smart pig run, Alyeska sent a janitorial pig south every four days for a month. When these pigs pop out in Valdez, they usually push out ten or twenty barrels of wax. In the pig mobile, they go straight to the pig wash. The wax, a hazardous material, is collected in barrels and shipped out of state. Once, not many years ago, after the pipeline wasn't pigged for six weeks, a pig pushed out forty-seven barrels of wax.

---

*TAPS has, though, leaked, and for the decade following two leaks caused by the settlement of the pipeline on June 10 and 15, 1979, Alyeska employees referred to the second week in June as Leak Week. The company's integrity manager always scheduled his vacations then.

Beneath all that wax, on account of corrosion, the one-billion-pound pipeline loses in the vicinity of ten pounds of steel a year: the same as an old Ford. Most of that metal loss is on the outside of the pipe, where it's buried. The inside is, well, nicely oiled. The exception is inside pump stations, where the pipe branches through valves and turbines. In deadlegs—hydraulic culs-de-sac, where oil sits stagnant—microbial-influenced corrosion is a threat.

If corrosion struck uniformly, such that the pipeline lost metal evenly and consistently, maintaining it would be vastly easier. After a thousand years, 99.999 percent of the pipe would still be there, sans weak spots. But rust doesn't work like that. It concentrates in relatively few places, begetting more rust. Alyeska responds only to those places that present severe integrity threats. It looks at spots where 35 percent or more of the pipe's wall thickness is gone, and where metal loss leaves the pipe at risk of bursting, which it determines from a formula developed by the American Society of Mechanical Engineers.* In pipeline lingo, this response is known as intervention criteria. Alyeska's are stricter than most. The company also gives itself a safety margin. It assumes that tiny pits are six inches long, and it assumes that the pipeline is operating at maximum pressure, when, in fact, as it runs out of oil, it is not. Depending upon where those spots are located—whether they're in sensitive or populated areas—Alyeska assigns each significant threat a PYTD: a Potential Years to Dig. The severest corrosion threats earn a 0 and get dealt with immediately. The mildest get 8s, or 15s, or even 29s. In this manner, even though Alyeska may address a dozen repairs a year, a backlog accrues. Examining the backlog, it is easy to be alarmed by not-infrequent reports of pits half the thickness of the pipeline.

Neogi exhibited no such alarm. "Take an average pipeline—Williams, Enbridge, BP—we're in pretty good shape," he said on the day he launched the smart pig. "TAPS was really well thought out. They designed a thing that'll stand the test of time." About maintaining the Alaska pipeline, Neogi made two analogies. "People think that a vehicle can last only ten years," he said, "but that's not true. If you have ten people working on it—"

---

*ASME's burst-pressure formula, called B31G, is more of a matrix. For a pipe of a given diameter and thickness, it reveals the size of a tolerable wound. On TAPS, for example, a pit 0.03 inches deep presents no hazard if it's less than two feet long. A pit 0.1 inches deep presents no hazard as long as it's less than a foot long.

He dropped it for a better one, comparing the pipeline to a painting at a museum, where curators, monitoring humidity and light, keep master-pieces pristine for hundreds of years. "The only thing that would stop us," said Neogi, "is the flow of oil . . ."

MILE 104

In every direction, all that was visible was white and blue. The North Slope was as flat as Kansas and just as treeless, but a good bit colder. It was cold enough to freeze pen ink. The only break in the flatness was a couple of stalagmitic ice mounds, called pingos, sticking up like erratics. The sun, low on the horizon, reflected off the frozen Sagavanirktok River and made mi-rages out of the distant Franklin Bluffs. The pipeline, zigzagging along, ap-peared inconsequentially small. Beside it was an orange snowcat, a treaded vehicle common at ski resorts, and inside, behind frosty windows, were two men, waiting for the pig.

Years ago, when pigs hustled along more rapidly, it was possible to hear them pass above ground or below. One employee, at Alyeska since the beginning, said pigs sounded like jets in the movie *Top Gun*: *sheeeeooow!* A man who didn't trust his ears as much could balance screws on the pipe-line—at least where it was above ground—and wait for the magnetic field of the pig to knock them over. Or he could jam a screwdriver into a part of a valve along the pipeline and listen to vibrations in the handle. Such be-havior didn't last long in the Arctic. Now these men listened for the smart pig with a geophone: essentially, a big, amplified stethoscope.

To do so, they strapped a metal probe onto the biggest appurtenance of the pipeline: in this case, the split ring of a flange. To block the sound of the wind, they wrapped the probe in a sheet of oil-absorbing cloth—what they called a sorbie—and duct-taped it in place. Wired to the probe was a battery-powered radio transmitter, which they set in the snow. Then, rather than stand out in the cold, they sat in their snowcat, turned the radio to 107.1 FM, and listened to the vibrations in the pipe while keeping their ears warm. They turned up the bass and turned down the treble. Since the FM dial was empty across the vast range of the North Slope, 107.1 broadcasted local static. When the men threw snowballs near the pipeline, the radio chirped. When they tromped around in the snow, it was as if

Indians were on the warpath. *Thunk thump thum thum thump.* They set up a half hour before the pig was due and then sat quietly, with the engine of their small machine turned off. In the giant Arctic, they were listening to a radio station that didn't exist, looking for something that you couldn't see, which in turn was hunting for something that didn't belong. As the pig approached, it produced a rhythmic beat as it regularly hit welds along the pipe. When the pig went by, it whooshed unmistakably, unlike any standard radio song.

When the smart pig passed by, it also stopped a clock on a toaster-oven-sized apparatus that picked up the signal from the pig's transmitters. This second, more exact method worked this time, but it doesn't always, even if the transmitter is the Alaska model. For some reason, it's less reliable than a pair of ears, and because the men didn't understand exactly how it worked, they didn't trust it. In any case, when the pig passed by, the men radioed Alyeska's Operations Control Center (OCC), relaying their identities, location, and the time the pig passed.

Listening carefully to those calls not far south was a young engineer named Ben Wasson. Neogi's lieutenant, Wasson was coordinating the pig run out in the field. (Neogi had since flown home to Anchorage.) Camped out in the cafeteria at Pump Station 3, Wasson was drinking coffee from a plastic cup and carrying around two handheld radios, two cell phones, and a logbook. During every call, he jotted down notes. He noted when another crew, twenty miles south of the first, reported hearing the scraper pig go by, and he noted when another crew, farther south still, reported having raised valves along the line, so that both pigs could proceed unobstructed. From these notes, Wasson compiled daily tracking updates, and these he emailed to eighty Alyeska executives, including Neogi. Attached to each update was a spreadsheet called "Pig Passage Summary." For all those antsy about the pig getting stuck in the pipeline, this document calmed their nerves. It demonstrated that the pig, surveilled every few miles, wasn't hung up or lost but on schedule.

Wasson was wearing Carhartt pants, a gray flannel plaid shirt, a black fleece jacket, and a green cap. A Mainer, from Bar Harbor, he'd shaved his lumberjack beard seven years ago, leaving a pair of rosy and windswept round cheeks. At the Formica table where he sat, he was joined by a grizzly bearded man who resembled a weary sailor. This was Dave Brown, the superintendent of the three field crews. Wearing a tan baseball hat, and resting his glasses on the second button of his blue T-shirt, he looked more the Mainer than Wasson. He sat down and began eating a hamburger. A NASCAR race unfolded on a TV nearby.

Brown has been on every one of Alyeska's pig runs since 1995, and on the first ten runs, after tracking pigs in something like 1,500 places, he only missed one. Hence his nickname: Super Dave. He later told me, "My crews don't miss pigs." It was his job to deal with any snags that came up. Earlier that morning, a snowcat en route to raise a valve at mile 81 broke through the ice of the Sagavanirktok River. Before it started sinking, the driver put it in reverse. He called Brown, who called Wasson, who called in a helicopter. Wasson had two helicopters at his disposal, as well as a front-end loader, two snowmobiles, and four satellite phones. When the radio system went down earlier, these proved handy. Neither Wasson nor Brown, though, had a spare snowcat. With one broken down, crews hadn't been able to track the scraper pig between mile 18 and mile 40. Had it gotten stuck, the smart pig would have slammed into it. Things worked out, but Brown remained unhappy. He didn't like missing pigs. After three days of near nonstop travel, he was also tired. After lunch, he took a nap.

With Neogi away, Wasson became the de facto guru of the pig run, its natural leader, choreographing the layers of logistics while drinking a lot of coffee. This pig run, in 2013, was his first since he started at Alyeska in 2010. He called the smart pig the ILI pig, for in-line inspection pig. More calls came in over the radio. Thanks to the tracking of Super Dave's crews, Wasson learned that the twelve-hour gap between the scraper pig and the smart pig was down to ten and a half hours. Wasson relayed the information to Neogi, who decided to delay the smart pig at Pump Station 3 for three hours.

Wasson is an engineer, but not the mechanical or hydraulic type. He's a civil engineer, and also a surveyor. He's a map guy, and thinks like one. The first date in his logbook was July 2, 2012—nine months earlier. One hundred and five pages bore notes on the pig run, all written in pencil. Wasson tapped the book, and said, "This won't break down. I can go without everything else." That was a surveyor talking. Thirty-seven years old, Wasson could have talked until he turned forty about surveying methods, precision, corrections. In a long discourse on digital mapping and Geographic

Information Systems (GIS) Wasson brought up the 2010 San Bruno, California, pipeline explosion. Pacific Gas and Electric, he said, knew that it had a deep pit, but thought the pit was on a thicker piece of pipe. "They had no integrity in their integrity management," he said.

———

Contrary to received wisdom on the lines, pig is not an acronym. It does not stand for pipeline inspection gauge. The Texas roughnecks who came up with the term in the early twentieth century did so after shoving bundles of barbed wire and straw through their lines. Apparently these bristly messes, which emerged covered in muck, brought swine to mind. Plus they screeched like pigs. Before Texans called them pigs, others in Pennsylvania called them go-devils, moles, rabbits, and spears. Further proof that *pipeline inspection gauge* is a crafty modern backronym lies in the term itself. Objects shoved through pipelines didn't do any inspecting until halfway through the century. Until then, they just scraped pipelines clean.

Almost any object can serve as a cleaning pig. Balled-up canvas rags and bundles of leather initially served the purpose. After a four-inch gas line in Montana was buried in a rock slide, an operator, in 1904, pumped a rubber ball through. One thrifty pipeline operator used chunks of foam mattresses. A jam manufacturer pigged his lines with loaves of bread. A paint manufacturer used plastic coffee cups. Tools engineered specifically for the purpose, more or less resembling encumbered plungers, have been available since 1892.

Some engineers have argued that the first smart pig was the first pig that got stuck in a pipe, but this seems like bestowing Phi Beta Kappa status on a preschooler. A failed cleaning does not an inspection make. The first pig to get stuck revealed that there was a problem but told nothing of the problem. If anything, it exacerbated it. On the other hand, unintelligent pigs did yield valuable information: some, made with soft aluminum discs or notched sheet metal, revealed evidence of encounters with dents, though not details about their locations. Tracking such pigs, even with chains or whistling noisemakers fastened to them, was difficult at best. Finding a stuck one was nearly impossible.

Then caliper pigs were devised. The first one had two fixed arms, sized differently, like a lobster, and a third articulated arm. Its motion, recorded on a chart, revealed where dents and dings were located. But it could go only so far, because it was powered by an electric cord trailing behind it. Not until 1955 was a battery-powered caliper pig designed. Not until 1959 did one really work.

In the meantime, much more capable pigs were conceived. Between 1953 and 1959, the field burst to life as nearly every type of modern pig was imagined. Four oil companies filed patents for pigs that could infer the thickness of a pipe based on magnetic flux leakage. If credit is due anyone, it is Howard EnDean, of Gulf Oil's Pittsburgh research lab, who filed four pig-related patents on one day in the summer of 1956. In addition to scheming up a leak-detecting pig based on sound and pressure, En-Dean designed an electrical-potential-gradient pig that would "ascertain the points where excessive corrosion is likely to occur." The only pig not patented during this tiny window was an ultrasonic thickness pig, the technology for which was a dozen years out of sight. Nevertheless, EnDean— the grandson of an old Pennsylvania oil driller—was a visionary. He was obsessed with the insides of pipes. Before he died in 1996, he expressed his wish to have his ashes poured into an oil well so that he could finally see what happened at the bottom.

Though the principles of his smart pigs made sense in theory, in practice they didn't work. Shell built the first MFL pig and soon declared it a failure. The tool was insensitive and could detect only corrosion on the bottom 90 degrees of a pipe. A subsequent pig developed by a company called Tuboscope could read all 360 degrees, but it could detect only major flaws, couldn't tell if they were internal or external, and it didn't have an odometer, so locating the pig's discoveries was a challenge. An MFL pig developed by Pipetronix in 1972 performed no better.

The major challenge in pig design lay in data storage. When British Gas, in the early 1970s, tested the MFL pigs on the market, it concluded that none was good enough and began developing its own. Storing the data, someone at British Gas said, was like "reading the Bible every six seconds." In an age of magnetic tape and paper charts, taking billions of measurements wasn't possible. As a result, early pigs were limited geographically, as if on parole, by their recorders. The Pipetronix pig could run only thirty miles, by which time its twelve-channel recorder had produced an inspection log a thousand feet long.

Pig engineers responded by designing pigs that collected different information. Instead of measuring metal thickness, they hunted for leaks. The best leak-detecting pig design relied on the detection of the inaudible, high-frequency squeal made by a liquid or gas escaping through a small hole in a pipe. Today this technique works marvelously. A generation ago, it did not. The sound of a pig traveling through a line was just too similar to the sound of a leak in that line.* And still, storing the data was a challenge. Shell made such a pig that could run for no more than four days. After that, it ran out of room to print. Sophisticated for its day, the recorder printed a seven-digit code every couple of seconds. The first digit indicated low or high pressure. The second digit indicated a leak, a marker, or no sound. The next five digits indicated the time, in hours, minutes, and seconds. To avoid duplication, the recorder printed out only the first two digits, up to nine times, until something changed. The result was a long, skinny ticker tape that looked like this:

1L24392 1L 1L 1L 1T24515 1T

Into the 1990s, TAPS was considered a "staggering inspection assignment" on account of its scale. In data storage and analysis, only recently have computers caught up. Modern pig data—color coded and laid out horizontally, as if the pipe had been split open—can be deciphered more easily. On TAPS, it's converted to a list of hundreds of thousands of calls and then winnowed to a short list, a few pages long, of anomalies that demand attention.

The second problem with early smart pigs was battery power. One had only enough juice to run for twenty hours. To save power, some tried to make pigs with generators that ran off wheels that spun as the pig traveled along, but the only way to keep the wheels from slipping was to give them teeth, which presented a catch-22. Designers tried to put a turbine in the center of a pig, so that fluid passing through would power it, but dirt, wax, and dust clogged it up every time. In an attempt to save power, more than one company designed a camera pig for use in gas lines. One company used a Hasselblad camera, the same as NASA had used on the moon

---

*That's why pipeline operators hired men to walk the length of pipelines and do vegetation surveys, hunting for clues that gas was leaking, and why others relied on dogs that sniffed out leaks and barked when they smelled something funny.

landings. Its lens captured the bottom 60 degrees of the pipe. The film was Kodak. The camera had no shutter, since a pipeline interior is blacker than space. Every forty feet, a strobe fired, and the camera got a shot. When the film was developed and printed, analysts used trigonometry to determine the scope of corrosion problems.

Getting a photo on the moon was easier than getting one in a pipeline. Dust obscured everything. One such pig, trailing behind a caliper pig, took photos whenever the caliper pig detected a dent. The good news was that it took surprisingly crisp photos. The bad news was that when the pig hit a dent, it shook so forcefully that the trailing camera overreacted and, according to two pipeline historians, took "excellent close-ups of the opposite but undented pipe wall."

An ultrasonic smart pig—which measured the thickness of pipe walls by listening to echoes—wasn't patented until 1971. It took many companies many years to develop this ultimate of measurement technologies. In early lab tests, ultrasonic pigs were very accurate, finding voids, inclusions, pits, and laminations. The hitch was getting them to work on dirty, rough pipes at twenty miles per hour. The first challenge was holding the transmitters and transducers firmly against the pipe wall in a "known, consistent relationship." The second was perfecting digital circuitry capable of measuring time in millionths of a second. This put ultrasonic pigs even with other varieties of smart pigs, whose history of runs in 1977 was, in the words of an engineer, "mixed." It was in that era that the Alaska pipeline began operating. As a result, Alyeska endured a good bit of pigging heartache.

⁓

Neogi chose Baker Hughes's tool, the Gemini, by first selecting the "threat categories"—the detection capabilities—of the pig he wanted to run. He wanted an MFL pig that could examine internal and external corrosion, particularly between the pipe and the seventy thousand clamps that hold the aboveground pipe to big stilts. The Trans-Alaska Pipeline traverses a great deal of permafrost, so about half of the pipe is elevated on H-shaped stilts. If the warm pipe were buried in these areas, the permafrost would melt, allowing the pipe to sink, bend, crack, and leak. The last time those regions had been examined was in 2001, back when Alyeska ran an ultrasonic pig. Neogi also wanted a pig that could detect dents and settlement. (Until 2009, Alyeska collected corrosion data and curvature/deformation data from two different pigs, in two different runs.)

From six bids, Neogi selected one from Baker Hughes, but until its pig was tested, or as Neogi put it, verified, the deal wasn't done. To verify the pig, Alyeska had shipped five pieces of pipe, laden with manufactured anomalies, and two clamps, to Calgary, Alberta. There, after the pipes were welded together in the gravel lot behind its building, Baker Hughes pulled its pig through. They call this a pull test, and according to Devin Gibbs, a serious thirty-five-year-old Baker Hughes employee with much pigging experience, it was a bitch. The pig weighed so much and had so much drag that the motor pulling it was maxed out, and the pig didn't want to budge. The thing was ornery then, too. But the pig found all the defects, Baker Hughes got the contract (a $2 million lease), and Alyeska was satisfied. Better: Neogi called this pig "a dream, a holy grail."

Still, the pig would be further calibrated. Alyeska has 150 intentional defects in its pipe at various pump stations. Once the pig emerged in Valdez with all of its data, analysts would use these to fine-tune the pig's defect-sensing algorithm.

In spite of Neogi's obsessing, there was one trait he could not guarantee in the pig of his dreams: he could not certify that this pig wouldn't get stuck.

In pipelines around the world, pigs get jammed in offtakes, wedged in valves that aren't entirely open, stalled at branches or wyes, constricted by debris, stuck nose down, trapped in reducers, pinned in too-tight bends, or—as on TAPS—sucked into drain lines. If one pig pushes on another pig in front of it, the added pressure can translate into increased outward force on the front pig's seals, stopping it like a cork. If the seals of a pig wear down or buckle, and product flows around it, the pig may stall in place, like a kayaker in an eddy.

A stuck pig may be jammed so tightly that the only way to remove it is by burning it away. Or a pig may get wedged so tightly in a sharp bend that, as happened in British Columbia, its batteries explode. A stuck pig has forced Canada's Keystone pipeline to shut down. In a subsea pipeline, a stuck pig is a particularly big deal. Runner-up is a stuck pig in the Arctic in winter.

Often what transpires between pig and pipeline is mysterious. Pigs pop out inside out. Cylindrical pigs pop out spherical. Some pop out backward. A thirty-six-inch foam pig has arrived inside another. Four thirty-six-inch steel pigs—each four feet long—got stuck in a train wreck and emerged 75 percent shortened. Most often, pigs pop out in shards and get swept up with a broom. Sometimes only the front half pops out, or nothing comes out at all. A smart pig has gotten stuck for nine months, and then all of a sudden "woken up" and finished its run. Conoco launched a six-inch pig on a thirty-mile pipeline in 1972. It popped out in 1996.

A pig can do worse than get stuck, though. Propelled by gas, a pig can easily rupture a pipeline. Behind compressed air, a pig has reached 170 miles per hour. A pig traveling at that speed, or a fraction of that speed, will refuse to take sharp turns. It will shoot straight through a pipe. A pig has created its own exit in a twenty-inch pipeline. Another has created its own exit at the bottom of a long downhill stretch. In late 1995, after a smart pig undergoing tests got stuck in Battelle's Ohio test loop, representatives of the pig manufacturer, thinking themselves wise, inserted a pusher pig behind the stuck pig, to force it out. This result was not achieved. Instead, the pig heated up, ignited a fire (the line held only air), and the test loop exploded. Both pigs were destroyed. Battelle—a nonprofit research company—later sued the pig manufacturer.

More than a couple of pigs have created their own exits at the end of the ride. A six-inch pig went through the door of a receiving trap and into a chain-link fence fifty yards away. A forty-two-inch pig came in with such velocity that it knocked a three-thousand-pound door three hundred yards into a car. A forty-eight-inch pig broke through a steel grate, slammed through a shed behind it, and landed in a pile of lumber. The first witness to the scene said it looked like a tornado had hit. A foam pig projected out of its receiver has cracked a brick wall thirty yards away. Another has cleared eight hundred yards. One made it half that distance, crashed through a wall, and landed in a bedroom in Texas. The homeowner said the room "looked like a war zone."

As such, a pig in a pipeline is like a cannonball in a giant cannon, and more than a few roughnecks have been killed—impaled or decapitated— by looking down the barrel. Trying to dislodge a forty-five-pound cleaning pig that had gotten stuck in the trap of a twelve-inch-wide pipeline, a New

Mexico man narrowly skirted death. He'd just opened the trap door when the pig, with 250 pounds per square inch behind it, struck him at ninety miles per hour. His skull and neck were fractured, his elbow dislocated, and his hand crushed. Examining the pig that came to rest three hundred feet away, investigators later found a two-inch piece of the man's arm bone lodged inside it. He survived.

That's why, at Alyeska, the crews were so slow and deliberate when loading and launching the pig. Once it was in the pipeline, though, fingers were crossed, because a lot could happen to the pig, as Alyeska has learned by experience.

On the Trans-Alaska Pipeline, pigs that avoided malfunction and escaped ingestion have gotten stuck at check valves and chomped by ball valves. In 1978 a pig got crushed by a valve at Pump 10 and came out in pieces in Valdez. The following year, a pig got stuck at a valve at mile 158 and wasn't removed for a month. A pig stuck at mile 15 on a Friday in 1984 made controllers so mad that they refused to hear about it—or deal with it—until after the weekend. Because of the refinery offtake in Valdez, pigs have come in with eight-inch bites taken out of them, as if attacked by sharks. Transmitters have been sucked off. In 2000 a pig passing through a valve at mile 467 took the valve's seat ring with it to mile 524, and a smart pig behind it pushed the ring from there to Valdez. In 2006 a scraper pig came apart in the line at Pump Station 7. Half of it was ingested at a relief valve there, and the other half continued down the line—becoming "the phantom pig." Since 1984, Alyeska has put a transmitter on every pig inserted in the line and done everything it can to keep every one from getting stuck.

MILE 144

Of all the pump stations along TAPS, Pump 4, nestled in a valley on the northern flank of the Brooks Range, is the most dramatic. At 3,200 feet, it's the highest pump station on the line, and more than any other, it felt like a ski lodge. The road in was frozen and washboarded and covered in drifts. Ridges on a peak to the northeast hinted of the Grand Jorasses. Pump 4 also distinguishes itself by two appurtenances in its manifold building. The station has a pig receiver and a pig launcher—making Pump 4, at least from a pig's point of view, the only true rest stop on the line. Between Prudhoe Bay and Valdez, only here would the pig take a break. This break would allow Neogi to confirm that the tool was performing as planned. On the morning of March 18, the pig made its way into Pump Station 4.

Neogi (who'd flown back to the North Slope) and his team were in the long, narrow building. It was dark, with little working space between the receiver and the launcher. From Baker Hughes, there was Devin Gibbs, who seemed to live in a thick blue jacket. He had run pigs in Mexico, Colombia, and Algeria. In a deep voice, he said, "That's the one country I want to go to: Australia. Check that one off the list." He recalled tracking a pig in a Korean gas line that ran under fields of rice paddies. The crickets sounded exactly like the equipment. "We had no idea where that pig was," he said. He was determined to succeed in tracking this pig across the North Slope, into Pump 4, and across the rest of Alaska. When the pig arrived at nine forty-five, Gibbs tracked it past the last valve and confirmed that it was squarely positioned in the receiver.

As with the launch at Pump 1, a dozen men in Tyvek suits and booties participated in the pig's receipt. They walked carefully on a floor lined with two layers of plastic and a layer of white oil-absorbent pads. They were prepared for the pig to emerge filthy. Among the crew was one new man: Baker Hughes's data analyst, a confident thirty-three-year-old named Matt Coghlan. He was there to confirm that the pig's sensors had worked and gathered the data.

Over a couple of hours, the Pump 4 techs closed the valves behind the pig, drained the trap, and opened the yoke on the end of the pipe. Then, slowly, they pulled out the tray with a winch. The pig lay on the tray. When Gibbs first saw it, he let out a big sigh. It was oily, but it was barely covered in wax, and he was greatly surprised and relieved. The pig was so clean that the other contractors called off their plan to steam wash it. Instead, rags and cans of brake cleaner were sufficient. Neogi called the pig's cleanliness a great success, and his boss agreed. He said it was cleaner than any pig run he'd ever been on.

The pig was also in fine shape. The cups, brushes, and sensor heads were no worse for wear. All appendages were present. In the tool's rolling ring, Gibbs discovered a three-quarter-inch ding and figured that the damage was sustained as the pig clambered into the trap. "Not a concern," he said. "We're still good." It was so good, he said, that he could have turned around and put the pig right back in the line.

Together, the Baker Hughes guys and the crane guys lifted up the pig, moved it, and placed it on a tray. The Pump 4 technicians then began closing the trap. Neogi didn't have an assigned duty per se. Within minutes, after a colleague unscrewed a small cap in the rear of the pig—pretty much the pig's rear end—Coghlan plugged in a USB cable. He connected the cable to his laptop and began checking the data. First he checked the pig's event file, to make sure the tool's systems had suffered no gross injuries. The file was small; nothing had gone wrong. It then took four hours to download the data.

Neogi spent many of those hours pacing. Wasson tried to keep him occupied and out of the way. That was his new logistical challenge. Once he got the data, Coghlan went nowhere without his laptop, bringing it to the cafeteria when he needed fuel, and Neogi trailed him like a junkie. Over Coghlan's shoulder Neogi peeked, bugging him, asking about the data. Just let me know anything, anything at all, Neogi said. He asked Coghlan to check in with him hourly.

As long as his laptop purred, Coghlan stayed awake. He stayed up all night, until, after breakfast the next morning, he had preliminary good news for Neogi: between Pump 1 and Pump 4, the pig had retrieved 100 percent data. Analyzing it would take months, but the data were there.

Relieved, Neogi and his team got some rest. Neogi flew home to Anchorage to spend time with his family. Wasson flew home too and went fishing. Meanwhile, the Baker Hughes team remained at Pump 4, sprucing up the pig. The section it had just completed was merely the warm-up. The next 656-mile segment would be twenty-five times the length of a normal run, two times longer than anything Gibbs had ever worked on, and full of unique hydraulic challenges. As Neogi put it, it would be the mother of all smart pig runs.

———◡———

The first time I met Neogi, I asked him if there was anything he'd ever prepared for so extensively as this pig run. He answered quickly: his fish tank. It took a month of listening to him tell stories, and actually seeing the tank, to learn exactly what he meant.

Neogi fell in love with fish in Lugazi, Uganda. On the northern shore of Lake Victoria, he caught his first and kept them in the bathtub. He was a little kid—a *Muindi*—and he was in Uganda because his father's work as a civil engineer had taken his family there. From Calcutta, the family had already uprooted to Bangladesh and then to Mombasa and Nairobi. Neogi grew accustomed to adapting.

In Lugazi, Neogi's mother insisted on sending Neogi to the best school in Uganda. That it was an all-girls school didn't matter. For three years, before his family moved on to Dubai, he remained, becoming a serious and thoughtful kid. By fifth grade, he'd learned that he could make friends through sports. So he got good at sports. He ran and played soccer and badminton, and has stuck with these to this day.

By the time his family relocated to Tanzania and Ethiopia, Neogi knew Bengali and Hindi, and had learned Luganda, Urdu, and English. In addition, he learned Swahili, or tried to. Throughout the variations in his education, only science and math remained constant.

His first year at University of Alaska, Fairbanks, where a relative taught, went swimmingly until his grades arrived. They were Cs and Ds. Because he was nearly acing tests, he hadn't bothered doing homework. A school counselor explained that in America, homework wasn't just practice but required. Then, on account of his low GPA, his scholarships were revoked. Because he couldn't afford any other way to get around, Neogi resorted to a bicycle. Six months into his biking project, someone stole his seat. He kept riding, seatless. When it was 40 below, it kept him warm. He rode through four Fairbanks winters.

He barely had food money. On the invitation of a friend, he attended a thesis presentation on silicon semiconductors where there was free pizza. He realized thesis presentations always began with food, so he started hunting for theses. On bulletin boards, he found announcements for presentations on engineering, biology, biochemistry, geophysics. He went, and ate, and always listened. Over four years—through college and into graduate school—Neogi attended 150 thesis presentations, more than any faculty member. Sometimes he was the only attendee. He'd have chips and salsa, some carrots, and sit there, learning.

He went through an aeronautics phase, a jet propulsion phase, and eliminated medicine from contention. He double majored in chemistry and mechanical engineering. He stayed for graduate degrees in mechanical and environmental engineering. Through the engineering department, he got a job doing research on fuel cells. Through the same department, he met the woman he would marry. Neogi had been teaching an air pollution class and saw her across the hall, smiling. Months later, he bumped into Bonnie at a party. She didn't seem interested. After midnight, he asked for a ride home; on the way there, he offered to repay her by buying her ice cream. They went to the only place in town that was open twenty-four hours, a Denny's—the northernmost in the world—and stayed until five in the morning. On their second date, he made Indian food at his place. She noticed Neogi's itty-bitty 1.5-gallon fish tank, containing guppies, and thought, "Okay, that's normal."

By the time they got married, eight months later, he'd upgraded to a 55-gallon tank, and then a 180-gallon tank. For their honeymoon, they went to Las Vegas and stayed at the Mirage. Neogi's wife picked the hotel because it had a 20,000-gallon aquarium in the lobby. Fifty feet long and eight feet wide, it held a thousand fish. Neogi loved it. He didn't gamble. He didn't drink, didn't smoke. Every evening, for an hour, he sat in front of the aquarium and stared at the fish: snappers, sharks, Australian harlequin tuskfish, a queen angelfish. The sparkling water soothed him. Twice he asked the front desk if he could see the filtration system, but was rebuffed.

Like most engineers and owners of spinning minds, Neogi needs to be busy and in control, and he struggles to relax. Water is all that releases him. Snow doesn't do it; he's not even a little skier. Nor does hunting. Fishing certainly doesn't. (He won't put a hook on the end of a line, but he will use a dip net.) He didn't drink at the Mirage because he's never drank. He doesn't allow himself to partake in anything he could become compulsive about: coffee, alcohol, cigarettes, fast cars. He'd never gotten a speeding ticket, but once, while talking about pigs, he got going fast enough for a cop to flash his lights at us. He admits that he's terrible at finding balance, and zooms in so far to most pursuits, like photography, that he gets lost. His only exception is fish. When it comes to fish, he has indulged.

Returning to Fairbanks, Neogi upgraded to a 240-gallon tank, a 270-gallon tank. A 600-gallon tank, and a 1,000-gallon tank, weren't far off. Neogi said watching fish is meditative; that it fights winter depression. Neogi loses himself in the water. When he heard that I had free time in

Alaska, he suggested I make my way to Seward, home of the Alaska Sea-Life Center, curator of three 100,000-plus-gallon aquariums, the biggest in the state. He knew the place well. But it's not just the fish or the water that he loves. It's managing the system. Keeping fish demands the range of his skills. He couldn't have picked more technically challenging pets.

While pursuing his graduate studies, Neogi received an invitation from Schlumberger, the oil services company, in Texas. South he went to check it out. North he returned, horrified of Houston. After a youth defined by relocations, he yearned to stay in Fairbanks. He thought about becoming a professor. He thought about work up on the Slope: two weeks on, two weeks off. In 2000 his advisor arranged an interview with Alyeska. He was hired as an engineer in Alyeska's integrity management group. He'd never thought about corrosion before.

He was supposed to start in Fairbanks on a springtime Monday. The Friday before, his boss-to-be called and told him he'd need to get up at five in the morning and drive to Valdez because a pig was coming in. It was an ultrasonic pig, made by NKK, the Japanese company that had milled the pipe comprising the pipeline. In Valdez, the Japanese crew bowed and called him "Bhaskarson." The pig came in covered in wax, having failed to collect sufficient data. That's how he started: with a bad pig run.

During his first years as an engineer and engineering coordinator and engineering advisor—before he began calling the pipeline his "baby," before he began spending more time worrying about it than about his wife and kids, before the pipeline consumed him—Neogi had enough stamina to play competitive badminton. He'd first played as a six-year-old in Bangladesh and had played on Uganda's national team. He played six days a week, sometimes twice a day, and competed in Miami, Boston, Chicago, San Diego. He taught his wife to play, and competed with her, the two of them making it to the semifinals of a 2005 tournament. He befriended Piotr Mazur, the Polish pro, and hosted him at his house. Mazur stayed for three years, and, competing as a men's doubles team, Mazur-Neogi briefly became the country's number three ranked team.

When major incidents happened on the pipeline, Neogi usually had a racket in his hand. On October 4, 2001, when Daniel Carson Lewis punctured the pipeline with a high-powered rifle, Neogi was playing badminton at the University of Alaska's Patty Center. His boss called, and Neogi ran over to the bleachers to answer his cell. He put down his racket, went to work, and stayed up all night. A year later, during a big earthquake that threatened the integrity of the pipeline, Neogi was in the Patty Center when his boss called.

While playing badminton in Hawaii, Neogi met Amar Bose, the Bengali entrepreneur and audio equipment tycoon. They became friends, bound with mutual respect. Neogi admires Bose's rejection of luxury despite his wealth, and calls Bose the smartest guy he knows. Neogi tried to get Bose into pigging, and Bose tried to get Neogi into sound research. He offered Neogi a salary of $250,000 to be his director of research, which would have put him in line for the presidency of the Bose Corporation. Neogi spent a couple of days thinking about it and then turned down the offer. He wants to get where he goes on his own. He wants to make his own way.

I once asked Neogi what he would do if he won the lottery. He said he'd quit work and just take classes forever. Evolution. Linguistics. Astrophysics. Bioengineering. Oceanography. Ichthyology.

After 2006, when BP spilled five thousand barrels of oil at Prudhoe Bay, a lot of people tried to hire Neogi, whose colleagues invariably describe him as "super intelligent" or a "genius." BP tried to get him as an integrity engineer. Conoco-Phillips tried to hire him, too. He was offered a job as a regulator. Every few months, offers came in—from Houston, Denver, Chicago, Edinburgh, Alaska. Thwarted, inquisitors often asked about other good candidates. Qualified piggers are hard to find.

Bhaskar and Bonnie had two children. They named their son Brij. Their daughter, born a couple of years later, they named Brianna. The license plate on Bonnie's van is B TEAM. Since having children, Neogi has stopped playing badminton so competitively and taken up soccer. He started and ran the soccer league in Fairbanks, and played on the team that won the state championship five times. The team is called the Rusty Buffalo.

The year 2011 was a rough one for Neogi. In January his mother, stricken with breast cancer, fell into a coma and died. Meanwhile, he was resigning himself to a life in Anchorage, where Alyeska wanted him. In March he drove his thousand-gallon fish tank down toward the coastal mudflats. He'd planned the trip thoroughly and figured his fish had enough oxygen for eight hours, the duration of the trip. Halfway to Anchorage, the differential in his Grand Cherokee went out. Ammonia built up in the tank. Oxygen fell. "I thought I had thought of everything," he recalled. He didn't have ammonia-absorbing pads or an oxygen tank. The trip ended up taking twenty hours. Of two dozen fish—some of which he'd had for a decade—only four survived. More than sorrow, he felt pain. For not being prepared, he blamed himself.

Reflecting on that year, Neogi said, "What makes someone successful is how they react to failures." He tries to study failures because he sees them as the best way to learn. In sports and school, with friends and fish, even pigs in the pipeline, the lesson has sunk in.

Buying a house in Anchorage, Neogi had specific and unique criteria. Could the floor handle the weight of an even bigger fish tank? Was there a room he could devote entirely to filtration? Could he install a sound barrier? On the hills of Anchorage's east side, he found just the place. Upstairs it's got white carpets, tall windows, a large deck with nice views. Downstairs the basement is more or less divided in two. One half contains a media room, a treadmill, a bed, and a large TV, where Neogi likes to watch the great Lionel Messi play soccer. The other half is devoted to fish.

Neogi's current fish tank holds 2,400 gallons. It's eighteen feet long, six feet wide, and three feet deep. Its acrylic walls are two inches thick. It took a crane to get it in his house. Outside of the aquariums in Seward, it's probably the largest fish tank in Alaska. In its crystalline water, a dozen tropical fish—yellow, black, and blue, worth thousands—flit about. Four pumps move 20,000 gallons of water an hour through it. To deal with that much water, Neogi built his own filters. These reside in a closet that Neogi has converted to a fish tank control room. It contains its own custom ventilation system but remains hot and humid. Behind a door in suburban Anchorage, 61 degrees north, conditions are downright Tallahasseean. A smaller tank of coral provides additional purification. Because gunk and sludge still clog the lines that lead from the tank to the filters, Neogi pigs them once a month. He uses two-inch foam pigs that his father-in-law sends up from Wyoming.

While he showed me the tank, a pump alarm chirped twice, because the resistance on the impeller was increasing. Neogi needed to take it out, soak it in vinegar, and reinstall it. But he already knew that, because his fish tank is remote controlled. Sensors in the tank monitor pH, oxygen level, specific gravity, ozone, dissolved oxygen, total dissolved solids, temperature. Other sensors measure power draw, humidity, lighting, and leaks. From his phone, Neogi can monitor and control his fish tank. He can adjust the lights. If the power goes out, a backup battery will take over, and he'll get an email. If the power remains out for more than a day, a backup generator will keep his tank running for a week. If a fish detector detects no moving fish, he'll be sent an email. If the temperature changes more than two degrees in two hours, he'll get an email. Thanks to a motion detector, he can tell if a caretaker is feeding his fish as much as he or she ought to be. A video camera monitors his fish tank in real time. In other words, Neogi has built a miniature simulacrum of a triple-redundant fail-safe pump station in his basement.

MILE 160

By the time the pig was launched on March 26, it had been cleaned and recalibrated, its ceramic sensors swapped with others less brittle, its batteries replaced. Neogi made sure that the technicians knew how much oil the pig demanded to get moving, and launched the pig on schedule at four in the morning. He timed the launch so that the pig was still twelve hours behind another scraper pig, and so that it would crest Atigun Pass at noon, and then—in daylight—face the major challenge of the run.

That challenge was the far side of Atigun Pass, down which a pig, unconstrained by oil—in what's known as slackline—could hit ninety miles per hour. At that speed, the sensors cannot gather data and can easily melt or crack, rendering the pig blind. If blindness ensued, Neogi would have to rerun the whole section from Pump 4 to Valdez. That would be a multimillion-dollar mistake. It was little consolation to Neogi that it had happened before. He didn't want a failed pig run on his watch.

Alyeska had been dealing with slackline at the bottom of Thompson Pass, just north of Valdez, for twenty years before residents in a neighborhood called Heiden View began complaining of earthquakes. The shaking woke them up at night. The pulses were caused by a torrent of oil crashing into a flat pool of oil at the bottom: a waterfall within the pipe. Engineers determined that the vibrations threatened the integrity of the pipe and that the induced fatigue had a small chance of cracking it. Alyeska responded by installing back-pressure control valves in Valdez that pushed the "slackline interface" 1,800 feet higher, but not quite all the way to the top of the pass. Over the Chugach Mountains, the oil still cascades almost a half mile down, like a truck with no brakes.

Atigun Pass, over the Brooks Range, is almost twice as high as Thompson Pass. Pulsing at the bottom compelled Alyeska to install a sleeve—essentially a metal cast—there in 2003. Because there are no back-pressure control valves to constrain the oil through the descent at Atigun, engineers at Alyeska rely on a complicated, concerted technique known as tightlining to guide the pig through.

The plan for Neogi's run, formed months earlier, was to stockpile crude in tanks north of the pass and push as much oil as possible through the line ahead of the pig. Then, so as not to overpressure the pipe on the south side of the pass, Alyeska would inject a drag-reducing chemical into the oil, divert the extra oil into other tanks, and send the rest down the line. By splitting the flow, Alyeska would put a bend in the hydraulic gradient at that pinch point. In other words, it would keep the pipeline from bursting. It was a tricky procedure, evaluated and refined by hydraulic engineers, tested virtually in Alyeska's pipeline simulator, scrubbed by operational engineers, assessed for hazards, and practiced ahead of time by Alyeska. Because the procedure entailed bringing the actual pressure so close to the maximum allowable pressure, the company sought permission to go ahead from the federal Pipeline and Hazardous Materials Safety Administration. PHMSA, a branch of the Department of Transportation, wants to ascertain the condition of the pipeline as much as Alyeska but not at the expense of prudence. The agency granted permission. It also asked to watch the procedure.

Early in the morning, Neogi sent the smart pig southbound. In an ultrasecure, unmarked, nearly windowless concrete building hundreds of miles to the south, one of Alyeska's operators watched carefully, on high alert. Operators there respond to emergencies much like pilots. They do not panic. In the building, regulators from PHMSA watched the same screens. The smart pig was represented by a little pink capsule labeled "030," for pig number 30. Shortly after noon, it crested Atigun Pass. One screen flashed brightly, indicating that the leak-detection system was offline. The flow, normally 560,000 barrels per day, more than doubled. On a large flat-screen displaying the hydraulic profile of the pipeline, a yellow line (actual pressure) bumped right up against a blue line (maximum allowable pressure).

Back in the Brooks Range, just below the pass, three crews were stationed at gate valves in case anything went wrong; in case the pipeline burst. Neogi remained at Pump 4, where he listened to updates over the radio.

Nothing went wrong. Shortly after one o'clock, Neogi called his boss and said, "We're good. We only went up to nine hundred psi." He was in good spirits. The pressure was right on the nose. The major hurdle between Prudhoe Bay and Valdez had been cleared in style. To celebrate, the operators and hydraulic engineers feasted on a tray of pulled pork.

⁓

Smart pigs weren't smart enough to find corrosion during the pipeline's first decade of operation. The first ones could detect only wall losses of 50 percent and, according to the engineer who ran them, "didn't work very well; they couldn't find anything." As another Alyeska engineer put it, "A high level of verification of corrosion signals to actual natural corrosion was not achieved." At least twice in the 1980s, pigs found buckles on the verge of becoming leaks and earned some respect. In two spots, the forty-eight-inch pipe was only forty-two inches across. Responding to a buckle under the Dietrich River, where the pipe had settled fifteen feet, Alyeska sent out crews immediately. It was 67 below. But by investigating settlement of the pipeline, which led to a couple of leaks, crews found corrosion. They knew their pigs were missing something.

In the spring of 1987, Alyeska ran its first high-resolution MFL pig, made by a Canadian company called International Pipeline Engineering. Along the line, it found a dozen potential anomalies. Alyeska dug up each site and found corrosion at every spot. The next year, Alyeska ran IPEL's pig again. Nineteen eighty-eight had been a banner year for TAPS. On one day in January, more oil made it through the line than on any other day in the line's history. In June Alyeska celebrated its tenth anniversary, having shipped five billion barrels of oil. No pigs got stuck that year. No corrosion repairs were made. No emergency shutdowns were necessary. The pig run was in the fall, and everything changed.

Reading thirteen miles of paper, analysts determined that the pig had found 241 potential anomalies. Field investigations verified two-thirds of them, revealing pits as big as quarters. Alarmed, Alyeska had IPEL re-analyze the data. The second analysis was worse: over 400 anomalies. "We thought we wouldn't have a problem with corrosion," Alyeska's pigging engineer at the time said, "but our world changed when the pigs told us otherwise."

Nineteen eighty-nine was rough from the outset. First, in January Alyeska dug up the pipe and installed thirty sleeves. Then, in March there was the *Exxon Valdez* oil tanker spill. Cleanup notwithstanding, it had little to do with Alyeska. If anything, it revealed the merits of shipment by pipelines over barges—but the result was increased scrutiny and ill will. In June Alyeska ran its first high-resolution ultrasonic pig. It was made by NKK, and it could measure 10 percent wall-thickness loss. The three-ton, red-cupped, titanium tool was five times more precise than IPEL's pig, accurate to the millimeter. It stored information in a "black box" of the type employed in aircraft. NKK had five engineers on its crew, and they brought gifts for Alyeska employees. They lived at Pump 1 for a month, and did calisthenics in the manifold building. They wore matching green helmets and uniforms, bloused where they entered their boots, and communicated by hand signals. Before loading their pig, they rolled up a Shinto prayer, put it in the tool, then got on their hands and knees and prayed to it. The Alyeska guys should have prayed too.

NKK's pig also found hundreds of anomalies. The total, now, was nearly a thousand. Investigating them, Alyeska found that nearly three-quarters of the pig's calls were, indeed, corrosion anomalies. Almost a third would need to be repaired. Corrosion damage was centered in two spots on opposite sides of Atigun Pass: to the north, under the Atigun River, and to the south, under the Chandalar Shelf. By August, Alyeska was weighing replacing nine miles of pipe under the Atigun River. That January, before it could do the big fix, the company dug up the pipe and installed eighty-six more sleeves in that section. That Alyeska did this in winter is proof that the damage was severe; that it ran NKK's pig again in the summer of 1990 is, too. An Alyeska spokesman told the Reuters news agency, "It is normal that pipes buried in the permafrost, in water and earth, are subject to some form of corrosion." So much for rustproof. Alyeska's engineering manager, Bob Howitt, reassured a writer from *Popular Mechanics*. "These are not the kinds of rust patterns that cause a total rupture," he said. "They're the kind that might cause a pinhole leak." That fall, Alyeska began replacing all nine miles of pipe in the upper Atigun River.

By then, Alyeska had dealt with the two final stings of 1989. First, the Department of Transportation's Office of Pipeline Safety told Alyeska that it planned to conduct inspections annually. "What's surprising to us is the degree of corrosion," the DOT's pipeline safety director, George Tenley, said after seeing Alyeska's corrosion reports. "I don't think anybody expected to see this much rust this soon." Then, in December, Alyeska billed Alaska for the cost of the corrosion repairs. That's the financial arrangement for a pipeline down the middle of the state: Alaska gets oil tariffs minus the operational and maintenance costs of the pipeline that delivers the oil. (Alyeska wanted to raise its delivery rates from $3 per barrel to almost $4 per barrel.) In that sense, the pipeline isn't just a cash cow, it's a cash whale. Flowing into state coffers, oil money dwarfs the cumulative contributions of gold, fish, and timber. Alaska, seeing nine digits in red on Alyeska's prefiled tariff sheet, did not like the size of the operational and maintenance costs. The state said, Wait: *replacement?* Why aren't we getting thirty years out of this pipeline? Alaska objected, citing operational "imprudence." It was the unlikeliest dispute imaginable, like the Israelites denouncing God for providing too little manna. The state sued: *Alaska v. Alyeska.*

Alaska's attorney general, Douglas Baily, was vocal about his displeasure. "They tell us they just got this state-of-the-art surveillance system, and that's why they found this," he told a reporter. "But we're not buying that story. This corrosion didn't just happen. They've had other technology to use." While Baily was in DC looking into the matter, he heard just the thing. NPR aired an interview with Kevin Garrity, the genial and patient public affairs chairman of NACE. Garrity, who was also a partner at a small firm called CC Technologies, understood corrosion. The AG got in touch, and Garrity was on a plane to Alaska the next day. He brought his colleagues Kurt Lawson and Neil Thompson with him.

For a few weeks, Garrity, Lawson, and Thompson evaluated whether or not the corrosion damage in Atigun was normal wear and tear. Their initial assessment was negative. It's worth noting that Garrity's company usually worked for oil companies. In fact, CC Technologies at the time was doing a lot of work for the American Gas Association and the Pipeline Research Council, which was funded by the same oil companies that owned TAPS. Agreeing to work for the state, he and his partners drew a line in the snow—and their other work evaporated. They became threats to the industry.

After the men presented their initial findings in Seattle, the Federal Energy Regulatory Commission allowed a full forensic investigation to proceed. For the next three years, Garrity, Lawson, and Thompson, with the help of nine more people, looked at corrosion in TAPS. They checked out the coating, tested the cathodic protection system, examined the passive anodic system, sampled the soils, the water, and the rust they found. Alyeskans, many of them bitter, didn't make it easy. Supervisors denied them access to dig sites. Others refused to cooperate. Alaska resorted to depositions. Garrity recalled that the situation was "quite contentious," and Lawson recalled that it was "pretty uncomfortable." Out along the pipe, they relied on Alyeska for food, shelter, communication—everything. They were, to say the least, vulnerable.

The corrosion guys formed a hypothesis about the cause of rust in the Atigun River area: sharp crushed limestone (rather than gravel) beneath the pipe was tearing through the high points of spirally wrapped tape, and because the tape happened to be made of polyethylene, current from the cathodic protection (CP) system couldn't get through it and protect the pipe. Water was then getting to the steel, and doing what water and steel and oxygen do.

Alyeska had wrapped the pipe in tape because the coating was imperfect and because the federal government told the company it could improvise. The coating, an epoxy called Scotchkote 202, had been introduced in 1965. (It's made from the same Epon resin as can coatings.) By 1972, it was apparent that it cracked in the cold, and by 1974, it was apparent that it had "disbondment problems." It didn't stick. Alyeska, committed to three million pounds of the stuff, sued its manufacturer, 3M, and settled out of court for $24 million. Because Alyeska wasn't about to scrape all of the stuff off, it asked the Department of Transportation and the Department of the Interior if it could wrap over it with two products, Royston Greenline and Raychem Arcticlad II. The departments okayed the move. "Rather than do the smart thing," said Lawson, "they

Band-Aided it. I think they did what they thought would work, but I don't think they put a lot of thought into it. The whole concept of taking a failing coating and wrapping it with something else is kind of mind boggling today."

Lawson recalled presenting their findings in a weeklong minitrial at the Watergate Hotel, but it might have been any DC conference room. Alyeska blamed saltwater that had started corroding the pipe the minute it was shipped by barge from Japan, years before it was coated, welded, and buried. Alyeska pointed to its world-renowned design, its high-for-the-industry standards, its state-of-the-art pigging program. Yes, Alyeska admitted, construction was rushed at the end. But recall the oil crisis of 1973 and 1974: the rationing, the shortages, the eight-hour lines at gas stations. The country wanted the pipeline in a hurry.

It took ten minutes for the mediator to make a decision: the state could sue. The mediator declared that to better oversee the pipeline, the various state and federal regulators would combine forces in one Joint Pipeline Office. And the mediator instructed Alyeska to hire CC Technologies for the next two years as technical experts. The corrosion guys went out and got drunk at the Dancing Crab.

Returning to Alaska, Garrity and crew changed their image. Alyeskans realized that they too wanted to see the pipeline operating for thirty more years. On TAPS, they focused on making the CP system testable; Alyeska put tens of millions of dollars into it. They did some probabilistic modeling, a lot of finite element analysis, and figured out where corrosion would manifest. They compared their predictions with data from pigs, and were spot on. Lawson claimed that Alyeska didn't need to run smart pigs anymore, but that's an exaggeration. Said Lawson, "There were repercussions, but the science was good and benefitted the industry." Their work had integrity.

CC Technologies survived, grew, and was sold to a giant Norwegian company called DNV. Garrity, who called his work on TAPS some of the best in his career, became the president of NACE. "I always tell people," he said, "if you had asked us to help you and figure out what's going on, we would have, but you didn't ask us. Now at the end of the day, all that research came back, and the pipeline turned out better." In the meantime, although Alyeska and Alaska signed a cooperative agreement declaring their intent to hunt down and battle corrosion, they both suffered. The Government Accounting Office accused regulators of complacency and Alyeska of making unjustified assurances about the prevalence of corrosion on the pipeline. Then the US Court of Appeals in Washington, DC, dismissed the case, and left Alaska with the bill. The state never got over it.

MILE 450

South of the Brooks Range, the pipeline weaves, above ground and below, through remote territory. Down the Dietrich River Valley—past the Chandalar Shelf—it passes under bands of broken yellow limestone cliffs. It passes Nutirwik Creek, Kuyuktuvuk Creek. It passes Sukakpak Mountain, looming like Nez Percé. It goes through gold territory, past Gold Creek, Nugget Creek, Prospect Creek, and Bonanza Creek. Mostly, it's wild country, with Sheep Creek, Wolf Pup Creek, Cow Creek, Moose Creek, Porcupine Creek, Bear Creek, Grayling Creek. Apparently it's Bigfoot territory, too. To the pipeline, though, ravens pose a greater threat. Ravens pick at the pipe's insulation, and then water gets in. Alyeska spent millions installing bands around the seams of the insulation, and the ravens persisted, outsmarting engineers.

The pipeline wanders past Coldfoot: a frozen truck stop more than town. It crosses an ice road. It crosses the Arctic Circle. It crosses swaths of burned patches, where stubby trees cover the land like obtuse cattails. To the west, the peaks resemble Colorado's Sawatch, broad and expansive above the tree line. Southern faces harbor poplars. Northern faces harbor only snow. Narrow valleys harbor spruce, and the flat valley bottom harbors red willows.

North of the Yukon, the pipeline traverses broad, barren lands, much like I-80 across Wyoming. Nearing the big river, the pipeline meanders through increasingly green forests, whose tall aspens evoke Colorado high country. Over the big river, the pipeline hangs from a bridge. For a hundred miles, as the pipeline nears Fairbanks, it snakes through rolling forested hills. Through all of this terrain, crews tracked the pig for six days. Neogi (back in Anchorage again) got updates. Then, as the pig slid into town, he flew up to Fairbanks.

Neogi showed up on an abnormally warm day wearing a silky green shirt that went incongruously with his brown pants, and no belt. He was, though, wearing a big, shiny watch. He seemed hurried, pushy—perhaps frazzled from the flight. He doesn't like flying. He still hadn't heard the pig go by.

He took his team out for lunch at his favorite Thai restaurant. There Neogi asked Wasson if he'd seen a news clip about the spill on Exxon-Mobil's Pegasus pipeline. Three days before, while the pig was at mile 325, the Pegasus pipeline leaked five thousand barrels in Mayflower, Arkansas. Eleven days before that, a Chevron pipeline in Utah leaked six hundred barrels. Neogi pays close attention to incidents on other pipelines. Since 2006, he's been on the piping code committee of the American Society of Mechanical Engineers. As such, he hears about pipeline accidents in Australia, Russia, India—where intervention criteria are much laxer. I asked him about the ramifications of the spill.

"You know, my ultimate job is to stop a leak," he said. "So it's good news if it didn't happen to us. It will bring more attention, more money, more technical development toward what we do. I wish it didn't have to come from a leak on a pipe, but that's the way it is." Then he said that it'd be a headache to do the root-cause analysis of the failure and make sure that Alyeska was not exposed to the same threat.

At the table, Neogi pulled up a video on his phone and passed it around. It was a clip of a superhydrophobic paint shedding water like a penguin. He said he wanted to test the stuff on Alyeska's pigs to see if it will prevent wax buildup and make cleaning them easier. He would like to do this by 2016, when the next smart pig will be run. Neogi had come to town to check in with his team, and in the way of support, he brought up some other pig-related modifications he wanted to make too. On Thompson Pass, he'd like to replace some of the pipe with thicker stuff, so that he can tightline the section. Or he'd like to install some new valves. He hadn't landed on a precise fix yet, but he was thinking about one. It will probably cost many millions. By then, Alyeska may be monitoring the pipeline with drones. But preventing a leak is still preferable to discovering one. What he was looking forward to most is the installation of a new pig launcher and receiver at Pump 9. The $30 million project is slated for 2015. The new facility would allow Neogi to do the pig run in three sections instead of two. His crews would be able to get more rest. As such, this mother of all pig runs was the last.

As the pig headed through Fairbanks, Alyeska's public relations team was ready. Often during pig runs, a rumor of a spill emerges, spurred by the flurry of Alyeska trucks running around at all hours. Phone calls and a blurb in the *Fairbanks Daily News-Miner* usually set things straight. This time, the pig clambered by unnoticed.

"The public has low tolerance for pipeline failures," Neogi said later. He pointed out that over a hundred people a day die in car accidents—few ever making national news—but when one person dies in a pipeline accident, we get hearings in Washington. Two days before the Pegasus spill, fourteen cars of a freight train derailed in Minnesota and spilled four hundred barrels of oil—and earned much smaller headlines. Neogi said, "No other department has our kind of accountability. A leak is easily a billion dollars." Actually, a leak could be five times that. "If I fail, the company's looking at over a billion dollars in problems: image, cleanup costs, respect in Alaska." He said, "If I don't do my job right, if I sleep on the job, it impacts the state, the owners, the industry, future drilling." To all concerned with the integrity of the Trans-Alaska Pipeline, that last part is a grave concern.

~~~~~~~

As the flow of oil through TAPS decreases, pigging will become drastically more difficult. Below 400,000 barrels per day, it will become impossible to tightline Atigun Pass, because there's only so much oil a controller can store in the tanks at Pump 1 before he runs out of emergency wiggle room. By then, the slack section on Atigun Pass will be over three miles long. Below 350,000 barrels per day, the "slippage factor" of a cleaning pig will prevent it from scraping the line effectively. With the bypass necessary to keep the wax ahead of it in a slurry, there won't be enough force to push the pig forward. Alyeska will also need to run them more frequently—as frequently as during this run's cleaning regimen—and this makes controllers nervous.

Meanwhile, by 2015, the small percentage of water entrained in the oil will drop out and begin flowing in a separate layer on the bottom of the line. Collecting at a dozen low spots, it could freeze. In so doing, it could disable check valves or halt pigs. At a flow rate of 400,000 barrels per day (expected by 2020), a pig arriving in Valdez could be pushing a slug of water one third of a mile long. Alyeska may need a new type of pig to push out the water, because water will also corrode the pipeline. Compounding matters, lower throughput will make it harder for controllers to detect leaks. But that's still not the worst of it.

At 8:16 a.m. on January 8, 2011, an employee at Pump 1 discovered oil in the basement of the booster pump building. It wasn't much, but it was coming from a pipe encased in concrete, which made fixing it, let alone assessing it, difficult. Alyeska shut down the pipeline and directed more than two hundred people to Prudhoe Bay to work on the problem pronto. A day passed. A second day passed, as did a third—and still the pipeline remained shut down.

As half of Alyeska's employees worked on the problem, many of them around the clock, the oil in the line running across the state cooled rapidly. The high temperature outside was very low. Like travelers stranded at airports, two cleaning pigs were trapped in the line: one near Fairbanks, and one at Thompson Pass. On the fourth day, with a repair still days away, Alyeska sought permission from PHMSA to temporarily restart the line. PHMSA granted the request. Alyeska restarted the line to stymie ice formation and wax deposition downstream of the pigs, because ice and wax could lead to cascading, devastating results. The pig near Fairbanks was safely trapped seventy miles south, between valves at Pump 8. The pig at Thompson Pass was trapped in Valdez and pulled out. Three days after the temporary restart, a week after Alyeska first shut down the line, engineers bypassed the leaky pipe and restarted the line for good. The operator who started it up remembers it clearly. She says it was the closest that TAPS had ever been to becoming an eight-hundred-mile-long Popsicle.

This is Alyeska's great fear, its "worst-case event." Declining throughput may necessitate frequent cleaning pigs, complex operating procedures, smarter and tougher pigs, and increased maintenance—but these are nothing compared with the seizure of the pipeline. North Slope crude gels at 15 degrees. It gets so thick that pumps can't push it. It becomes thixotropic, like quicksand. For whatever reason—a power outage, say—if the oil sits in the line too long, at the wrong time of year, the threat of the big Popsicle looms. In January 2011, the oil cooled to 25 degrees. The threat is critical.

Alyeska's former president told Congress that at the flow rate expected in 2015, nine winter days of shutdown could spell the ultimate end of the pipeline. If the oil gels, there will be no recovering from it. The threat makes explosions and even leaks seem trivial. It's a game ender.

It's because of this conundrum that drilling in the Beaufort and Chukchi Seas is of such importance to Alaska, Alyeska, and Alaskans. Those rigs will tie into the Alaska pipeline, feed it their oil. Sure, residents will get annual dividends, and Alaska will receive billions in royalties and taxes that fund pretty much everything in the state. But it's the long-term future of the state on the table. The sooner that someone turns around the two-decade saga of declining throughput, keeping the pipeline from turning into a giant Popsicle, the easier those concerned with the integrity of the pipe will sleep.

In the meantime, if TAPS leaks for some reason, and the public withholds forgiveness, the resultant delay in offshore drilling could portend the end of the line. That's what Neogi was implying when he mentioned the impact on future drilling. A big spill could delay offshore drilling in the Beaufort or Chukchi Seas for two decades, and this could spell the end of the line. End of the line would be the end of the state of Alaska, and not exactly beneficial to the economy of the other forty-nine states in the union. Precarious is the future of the pipeline, and high are the stakes in which Neogi and the integrity management crew operate.

On the day before the pig left Pump Station 1, Secretary of the Interior Ken Salazar said that Shell "screwed up" in its attempt to explore off the northern coast of Alaska in 2012. While the pig was at Pump 4, Shell's offshore rig, the Kulluk, was secured for its journey across the Pacific Ocean to be repaired. Before the pig left Pump 4, Alaska's state senate voted 11 to 9 to cut taxes on the oil industry—billions over the next decade—to make future production more attractive. Without giving it away for free, the senators made the North Slope a more desirable place to drill. The subtext of all of it was the integrity of the Trans-Alaska Pipeline, without which nobody would have any way to get the oil.

Alyeska used to be shy about the challenges it faced. Now it's revealing them like battle scars, telling its stories to anyone who will listen. The company is asking for sympathy, understanding, and help. The CEO granted me unprecedented access. During my visit, he met with the *News-Miner* to talk about the declining throughput problem, mentioning a new "very low flow" study. The company's low-flow study is already posted on its website. Presenting to a DC think tank, he employed a creative analogy. Flow rates below seven hundred thousand barrels per day he compared with a "dipstick below the add-oil line." Everyone knows better than to drive a car with no oil. The rate hasn't been above seven hundred thousand barrels per day for years.

Every Alyeska employee—whether he or she prefers Rachel Maddow to Bill O'Reilly—wants more oil production. Federal regulators in Alaska feel the same way: they're not pipeline foes, and certainly not oil foes, and they aren't recalcitrant regarding this bias. They're pipeline boosters, tending to it and rooting for it so long as its operator adheres to minimum standards. All worry that Russian oil companies will get offshore in the Arctic before US companies do, and resent that politicians are stopping them from sinking wells in the Arctic National Wildlife Refuge, below which there's enough oil to triple the throughput of the pipeline for a generation. At least one regulator wasn't sure if the pipeline would be around when I'm old enough to collect Social Security. Todd Church, the manager of the operators in the unmarked concrete Anchorage building, said he foresaw the challenges of declining throughput when he accepted his job but figured there was enough oil to last his career.

Until Alyeska gets more oil, it's preparing for the worst. During the January 2011 shutdown, Pump Station 7 came to the rescue by heating the oil in the line. Now Alyeska wants to add heat to the line, at great expense, at four more pump stations. The company would like to install more insulation on sections of elevated pipe. It would like as much oil as it can possibly get from ConocoPhillips's new Alpine fields, and not long ago, it looked at blending crude with liquefied gas, of which there's plenty (thirty-eight trillion cubic feet) under the Slope. The authors of a study on the subject, from the Petroleum Development Laboratory, in Fairbanks, found that adding liquefied gas to crude raised the wax appearance temperature, and that the addition of liquefied gas disrupted the stability of crude, promoting significant amounts of asphaltene precipitation and deposition. Covering the walls of the pipe with wax was one thing; covering it with asphalt was another. Asphalt was "very undesirable," and though the authors never explicitly said why, the reason ought to now be clear.

They figured the next thing to do was study the addition of asphaltene-stabilizing chemicals.

Alyeska employees insist that the real question is not mechanical (How low can the flow get?) but financial (When will the producers decide it's not worth it to keep drilling on the North Slope?). Instead of throwing money at the challenges before them, the producers will seek opportunities elsewhere. Their financial obligations are to their shareholders, after all. Until then, the newest financial twist in Alaska's petro-colonial soap opera is the state's latest lawsuit against Alyeska. The state objects to the billion-plus dollars that Alyeska has spent on "strategic reconfiguration," which is jargon for low-flow modifications. As always, every dollar spent on the line is a dollar not going into Alaska's coffers. The way Alyeska sees it, there'd be no more pipeline if it weren't for those modifications. The weeklong shutdown in January 2011 would have been the final blow.

MILE 549

South of Fairbanks, conditions were so good, and the access roads to the pipeline so well plowed, that the crews tracking the pig didn't need snow-cats. Trucks sufficed. In North Pole, where the pipeline passes beneath a suburban development, tracking the pig was downright casual. Across from a small ranch house, Ben Wasson parked on a snowy road, and after setting up the geophone, he hung out in the road with the truck door open. A resident in a pink fleece jacket and a wool hat walked by with a golden retriever. Had her dog pooped right then, Wasson would have heard it on the radio. Over the next week, Wasson found himself coasting. His updates to Neogi and executives at Alyeska became carbon copies: "Tracking continues day and night . . . OCC continues to provide excellent support."

Near mile 500, where the pipeline wanders far from any road, tracking crews gave up. As a result, for a half day, nobody knew where the smart pig was. Everyone hoped it wasn't stuck. It was the day *after* April Fools' Day. The pig was fine.

By the time the smart pig was at mile 549, it was only nine hours behind the urethane cleaning pig. Engineers in Valdez wanted more wiggle room between the two pigs, because in January 2000 they learned the hard way that the pig trap there was not capable of receiving two pigs concurrently. The second pig was forced into an eighteen-inch line, as if made of Play-Doh. Engineers in Valdez also didn't want the smart pig jetting down Thompson Pass just as the cleaning pig arrived in Valdez, because a pressure wave, they feared, could trigger the relief valves and lead to the ingestion of the cleaning pig. So they decided to put eighteen hours between them. That meant holding the smart pig at Pump Station 9 for a half day. This scrambled Wasson's schedule and didn't thrill Neogi. In that larger window, he feared, the likelihood of wax buildup was greater. He hoped it wouldn't make a difference.

While the pig waited at Pump 9, twenty cars of a freight train in Ontario derailed and spilled four hundred barrels of oil. In San Francisco, opponents of the Keystone XL pipeline chanted, "When I say pipeline, you say kill! Pipeline! Kill! Pipeline! Kill!" In Alaska, such language would get you knocked out and left for a grizzly snack.

Two days later, the pig journeyed up and over Isabel Pass, through which the pipeline crosses the Alaska Range. The pass, relatively low and broad, wasn't nearly as tricky to tightline as Atigun. The pig slid through without incident on April 5. It was, by then, three-quarters of the way to Valdez. That same day, a Shell pipeline in West Columbia, Texas, spilled seven hundred barrels.

⌒⌒

Throughout the 1990s, Alyeska's pigs and pigging techniques smartened up. NKK's second-generation ultrasonic pig was loaded with 512 transducers that could each take 625 measurements per second. With that tool, Alyeska found and fixed nearly all of the gouges, dents, and deformations inflicted during the construction of the pipeline—what's known as first-party damage. With that tool, in 1994, Alyeska also found corrosion in 300 joints on the pipeline. When Alyeska informed the Department of Transportation of its findings, the DOT told Alyeska to run pigs annually until it got a handle on things. In the *Federal Register*, the official journal of the US government, this insulting statement appeared: "the design has not prevented all corrosion from occurring." Alyeska obliged and ran an ultrasonic pig in 1996, 1997, and 1998.

In 1996 Alyeska began storing all of its pig data in a massive single database. It gave corrosion engineers the opportunity to compare corrosion growth over time—to see if corrosion was active or passive. It also allowed company engineers to compare the reporting characteristics of its various pigs. Alyeska's integrity manager, Elden Johnson, and a few other corrosion engineers, including a PhD named Stephen Rust, presented a paper at NACE in 1996 that compared the performances between Alyeska's ultrasonic pig and its magnetic flux pig on runs between 1991 and 1994. Statistically speaking, the clear winner was the ultrasonic pig. But it remained far from perfect. It missed sharp edges, deformations, and wrinkles. In 1998 the pig reported calls at mile 710. Alyeska dug up the spot in the spring of 2000, found five gouges in the pipe caused during construction, and then discovered that one of them was 80 percent of the way through the pipe's steel. The pig had caught the anomaly but underestimated its severity. The Joint Pipeline Office was not pleased. It called corrosion control a "significant concern" and "a significant maintenance challenge."

Between 1994 and 1999, Alyeska excavated 165 spots that its pigs had alerted it to. Only a handful required repair. Compared to the blizzard of work at Atigun, this was a bunch of flurries. Based on its good record, Alyeska asked the DOT if it could pig less often: every third year rather than annually. The DOT said fine and told the company to resume pigging in 2001.

Then declining throughput began to affect the pig runs. The first failed run—the first which had to be rerun—was in the spring of 2001, on Neogi's first day of work. The pig, covered in wax, hadn't gathered enough data. To give it the best possible chance of a successful second run, controllers stockpiled the lightest, warmest crude and ran the pig in it. That rerun was the first full pig run that Neogi saw, and it was Alyeska's last dalliance with an ultrasonic pig. Henceforth, Alyeska couldn't get its pipeline wax-free enough.

Resorting to an MFL pig, Alyeska suffered the same setback in March 2004. Wax ruined the run. After running a bunch of cleaning pigs, Alyeska ran the MFL pig again two months later and got enough data. Alyeska discovered thirty-five significant spots that the 1998 and 2001 pigs hadn't picked up on at all, some of which had wall losses from a quarter to a third of the pipe. In particular, Alyeska found four dents with metal loss, one of which required immediate repair. Alyeska, though, neglected to inform PHMSA and didn't fix it for a year. To assuage PHMSA, Alyeska reassured regulators through at least the fall of 2005 that it planned to employ an ultrasonic pig on its next run.*

Four months later, in March 2006, external events screwed up things again. Severe corrosion in one of BP's North Slope feeder lines caused a quarter-inch hole, and through it, five thousand barrels of crude spilled onto the tundra. Where Alaskans used to joke that BP stood for Big Provider, now they said it stood for Broken Pipeline, or Bad People, or even Bureaucratic Pandemonium. An Alyeska employee told me that BP was running a sprinkler system rather than a pipeline system. As with the *Exxon Valdez*, Alyeska had nothing to do with the goof-up, but it felt the ramification: heightened attention to corrosion. Under the spotlight, Alyeska decided to send a smart pig through TAPS immediately.

That summer, Alyeska had just celebrated its thirtieth anniversary, having shipped more than 15 billion barrels of oil. That was 50 percent beyond what anyone had expected from the pipeline. Confidence outside the engineering department must have been high counting down to the pig run. Alyeska ran the pig in August. Under pressure to pig as quickly as possible, the company hadn't prepared a cleaning regimen. In fact, Alyeska hadn't cleaned the pipeline for most of March, following the discovery of BP's leak. Because the BP field was shut down, throughput in the pipeline was only 450,000 barrels per day—lower than it was in March 2013. With the low flow, wax buildup must have been rapid. It ruined the run. In Valdez, techs spent a week removing wax from the pig and discovered that 20 percent of the pipe had gone unexamined.

Alyeska ran the pig again in September. That time, in slack conditions, the pig scooted down Atigun Pass and got going so fast that the sensor heads broke when they hit welds. The pig stopped collecting data near the bottom of the pass. In March 2007 Alyeska ran the rebuilt pig for the third time—and got just enough data so that Neogi was able to combine what the pigs had gathered into a tolerable analysis.

The next month, PHMSA fined Alyeska $260,000 for analyzing the data from the second 2004 run too slowly. The law gave Alyeska six months after an integrity assessment to determine threats to its pipeline,

---

*Regulators at PHMSA now prefer that pipeline integrity be assessed from *both* magnetic flux pigs and ultrasonic pigs, the combination of which leaves no blind spots.

and the company had taken eleven. Alyeska objected, citing "technical difficulties" and electronics failures caused by wax. PHMSA lowered the fine
to $173,000. Still, it was the largest civil penalty levied against Alyeska in
a decade.

The architect of the failed pig runs in 2001, 2004, and 2006–07 was a
large churchgoing engineer named Dave Hackney. He'd come to Alaska
from New York in 1962, studied mining engineering in Fairbanks, and
helped build the pipeline. In 1983, from Alyeska's quality assurance
department, he volunteered to oversee smart pigging when nobody else
did—thinking it was good job security. For years he wore a hard hat
that said "Swineherd." Like many fastidious engineers, he maintained a
Teddy Roosevelt mustache and kept a stable of pens in his shirt pocket.
He claims to have taught Neogi much of what he knows. At Alyeska
now, his name is verboten, and his last runs are infamous. "Previous ILI
engineer" was how people referred to Hackney. In 2008, after Alyeska
brought in a new president, Hackney was transferred out of the pigging
group and soon thereafter left the company. He was sixty-two. "The official line is 'I retired,'" he told me. He sued his former employer and
settled the dispute out of court. Now neither party can criticize the other
or get into the matter. Hackney laid blame on tough circumstances and
"the politicians that won't let us drill for more oil," and insisted that on
TAPS, pig reruns aren't failures (millions of dollars notwithstanding) so
much as par for the course—and that the result is still good corrosion
data. He left it at "Whatever happened, happened." He remains proud of
his work defending the pipeline.

Neogi had some sympathy for Hackney, without going so far as to call
him a victim. "Who knew there'd be this much wax?" he said. "It's not like
we ever half-assed it." At the same time, he didn't think events concluded
in such a crazy manner. "It's a big deal. You wanna make sure you're taking care of the integrity of the pipeline. The definition of insanity is doing
the same thing again and again, and expecting different results. If I do this
three times and I fail, I hope Alyeska sees fit to bring in someone who has
better ideas. I don't think it's personal. It's an important task, and Alyeska
wants to make sure it gets done right. Dave, Elden, myself—we're replaceable. The pigging is not.

"In 2006, we learned a lot," Neogi added. Alyeska learned to run a much tighter ship, and Neogi learned not to hurry the process. It's the same thing he learned with his fish.

MILE 800

Twenty-four hours before the tool was due in Valdez, Neogi left Anchorage in his wife's minivan, the one with the B TEAM plates. He drove to Valdez not just because he hates flying but also because a storm across most of south central Alaska had grounded flights into Valdez. He had to get there, so he was driving three hundred miles through whatever awaited. Conditions on the first hundred miles of the Glenn Highway were bad. The Matanuska Glacier and the dramatic peaks of the Chugach, off to the right, were obscured in white. It took three hours to get to Eureka Summit, where conditions improved. Here Neogi—who'd been snacking on dried fruit—stopped for lunch. He was wearing a red T-shirt, blue jacket, green track pants, and sneakers. He was sporting a gold chain, the big watch, and two rings. He looked like he was on vacation, and not in Alaska. From two snow machiners at a nearby table, he drew a couple of looks.

Neogi sat down, ignored the bottomless twenty-five-cent coffee, and ordered a Eureka burger, pronouncing it "you-*wreck*-uh." He checked his BlackBerry, and had received twenty-five emails. One informed him that the final cleaning pig had recently arrived in Valdez. It was there, but it was still in the receiver. He called an engineer on his team and said, "Call me when the sweep tool comes out, and let me know how much wax there is." After eating quickly, Neogi bought a pack of gum. For him, it was the equivalent of a pack of cigarettes.

Sixty miles down the road, in Glenallen, Neogi took a right on the Richardson Highway, and paralleled the pipeline as he headed south. As he drove, he evaluated the pig run and considered the various hiccups encountered: the stuck valves, the helicopters, the tracking, the flow variation. He said that the most significant was the current twenty-one-hour separation between the sweep pig and the smart pig. He didn't like the possibility between the two for wax buildup. He held the index finger and thumb of his left hand a hair-width apart and said, "If the wax even builds up only one millimeter . . ."

He restarted. "I would like to have had more oil in the pipe. Quantifying wax is very difficult. There's no pure science that says this is it."

Then he said something surprising: "Even in engineering, there's things that are half magic, half repeatable."

Beyond mile 700, it started to snow again. Neogi insisted that he was relaxed. He insisted that even if the smart pig did hit a check valve, the pig's cups, rather than its electronics, would take the blow. He insisted that the pig wouldn't find any corrosion below the clamps on the stilts. "You just don't know," he said. "You do the best you can with what you know. I don't worry about it, because I've done my homework. There might be a little hiccup here or there. My game plan has always been to study the hell out of the thing. Where can you get better? If something happens that you haven't thought of, then you just put your hands up and say, 'Well, this one got by me.'" Ten minutes later, after a few glimpses of the Wrangell Mountains poking out from clouds to the east, Neogi asked me if a cold front had come in. Cold wasn't necessarily bad news, because frozen soil conducts less heat from buried pipe than does wet soil. Either way, at the start of the pig run, Neogi was thinking about delta P. Now, approaching the end, he was thinking about delta T.

Eighty miles from Valdez, as he was descending to the Little Tonsina River, Wasson called. It was 2:47 p.m. Neogi pulled over and put on flashers to talk to him. Neogi's half of the conversation went like this:

"Hey, Ben. You got five barrels of wax? . . . Yeah, but that seems high . . . What did the one before bring in? . . . Is it hard wax or soft wax? . . . Hmm . . . Well . . . You don't have any pictures? . . . Okay . . . Okay . . . Okay."

After Neogi hung up, he looked in the rearview mirror, pulled a little farther off the shoulder, and called his boss to report the news. When he hung up, he said, "Well, there's a surprise." A couple of barrels, he admitted, was what he was hoping for. Then he qualified it. "Five's not bad—it's not like forty," he said. Shortly thereafter, he approached Thompson Pass, where oil in the Trans-Alaska Pipeline cascades almost a half mile down the Chugach en route to the sea. Down the stretch, recent pigs have not been able to capture any data, and Neogi hoped the pig wouldn't exceed fifty miles per hour. As he neared the pass, he started humming Beethoven. It was the only other time I heard him do it.

After checking in to his hotel room in Valdez, Neogi spent the evening watching the NCAA basketball championship game. Then he had dinner at the only Thai restaurant in town. It was also the only place with an aquarium full of fish. Over dinner, he insisted that he'd sleep okay, that he had learned to deal with butterflies in his stomach long ago in badminton competitions, and that the last time he experienced the feeling was when his kids were born. He said that he knew that he had the most to lose but that he was also well prepared. He said it was like being in sight of the finish line at the end of a marathon. Crossing the line is all he wanted to do. He denied that crossing it would be exciting. "You can't be in this business if you're looking for excitement," he said. "Think of a war general: he can't be running around, going nuts at every little thing. He has to be calm and collected." In that manner, Neogi said he'd return to his room, call his wife and kids, read something, and—twelve hours before the pig arrived—sleep just fine.

Overnight, a foot of snow fell, and in the morning it was still snowing. The Valdez Marine Terminal may be the northernmost ice-free port on the Pacific, but it occupies one thousand of the most precarious and snowiest acres anywhere. Snow rose to the height of stop signs, and higher. Berms and snowbanks were so steep and gnarled in places that they resembled seracs; the streets veritable crevasses. One winter brought forty-six feet. The immense depth was more astonishing than the immense width of the North Slope. Sliding off of the roof, snow somehow broke a window on the fourth floor of Alyeska's modern glass office building. The famous statue honoring the (mostly) men who built the pipeline, with the plaque that reads "We didn't know it couldn't be done," was so covered in snow that it couldn't be seen and couldn't be gotten to. All over the terminal that morning, loaders and plows were crawling and beeping. In front of the building, men were shoveling.

Inside the building, Devin Gibbs sat in the cafeteria, ensconced in his huge blue parka, eating breakfast alone. It was his thirty-sixth birthday. The pig was due in an hour. He said, "The best birthday present would be if the pig comes in clean." Two flights up, Scott Hicks, the trim, affable director of the terminal, was walking the halls. BlackBerry in hand, gray V-neck sweater over a blue-collared shirt, he said, "I just hope it comes in in one piece." Behind wire-rim glasses, he winked. Ten minutes later,

Dave Benes, the lead technician who was about to remove the pig from the pipe, insisted that the action would be standard operating procedure. A big guy in denim coveralls, he had a confident mustache and goatee beneath a camouflage Alyeska cap. "Yard 'er out, get ready for the next one," he said. Neogi, who appeared preoccupied, said very little.

The pig receiver was just up the hill, in an innocuous tan warehouse called the East Metering Building. On the north wall of the building, there was a tall bay door and a rusty human-sized door beside it. A sign reading DANGER: MEN WORKING OVERHEAD had been duct-taped above two others that said KEEP THIS DOOR CLOSED and HARD HATS REQUIRED. On the east wall of the building, about ten feet up, just poking out above a snow bank, there was a little orange milepost sign. It said 800. The end of the pipeline was just inside.

At 9:05 a.m., the bay door rolled up. Inside, some familiar features were visible: a big yellow overhead crane, the little red gantry crane. Below them was the end of the pipe. Aside from THE END stenciled onto it, the end of the pipe didn't quite look like the end of a pipe—because four smaller branches protruded from the stub. They took the oil to the terminal's eighteen enormous tanks, from which it could be loaded into thousand-foot tankers at the berths down the hill. But that stub was it: the end of the line. In front of the door on the end of the stub, taped to the concrete floor, were two long layers of white oil-absorbing cloth. Beside the white carpet were black drums, lined with plastic bags—for shoveling wax into. There were eight of them. Beneath the floor, lodged somewhere in a section of concrete-encased pipe, was half of a pig that had been ingested in 2012. It had since been named Theodore.

At 9:10 a.m., Neogi, Wasson, and Gibbs walked in from the east door. Ten minutes later, Benes and three other station techs, in matching brown coveralls, hard hats, steel-toed boots, and gloves, entered through the bay door. One had a wrench in his pocket. Another had a radio strapped to his chest. The pig was less than a mile away.

It was April 9, 2013—twenty-six days since the pig was launched from Pump 1. During this span, the length of a day in Valdez had grown by nearly two and a half hours.

At 9:47 a.m., the pig arrived.

Last time anyone saw it, the pig was sixteen feet long. Nobody knew if it had been mangled or compressed. The pig trap was just under forty feet long, but the final six-foot section was a deadleg. To receive the pig, the techs planned to stop it in the thirty-four feet of the trap through which oil flowed. With the pig parked there, they could use the oil flowing past it to flush wax off of it. Then they would divert the oil, close the valves, drain the trap, open it, and yank out the pig. Between the trap door and the massive ball valve at the upstream end, they didn't have much room.

To get the pig to come in slowly (they didn't want to slam the $2 million device into the trap door), the techs slowly opened another gate valve at the far end of the building. As the valve began to open, oil behind the pig found an alternate route to the four tank lines, leaving less oil to propel the pig forward. Tracking the pig through the final hundred feet of pipe was easy. Screws placed on the pipe fell as the pig passed. Geophones caught the sound of the pig's movement. Receivers picked up the signals from the transmitters in the pig. Then the pig passed the ball valve, entered the receiver—which was really a pipe within a pipe—and tracking the pig became much more difficult. The screws no longer responded, because there were now two inches of steel between them and the magnets on the pig. The signals from the transmitters vanished. The geophones revealed nothing. The techs opened the gate valve a little bit more, until it was a quarter open, and enough oil took the new path that the pig came to rest. Where was anybody's guess.

Gibbs checked for transmitter signals, but they were null. He felt blind. It was the most difficult pig trap he'd ever seen, and he'd seen thousands. His only useful tool was an old-fashioned gaussometer. Registering the magnetic fields around the pig's magnets, it revealed the polarity switch between the two banks, giving Gibbs a six-foot window of precision. Wasson, ever the surveyor, used a plastic Boy Scout compass to the same effect. He and Gibbs reached the same conclusion: the pig had come to rest closer to the ball valve than to the deadleg. Gibbs was pretty sure the pig was clear of the valve, but to be absolutely sure, he needed three more feet. "You don't want your tail hangin' in the valve," he explained later. He'd never shut a valve on a pig, and wasn't about to by making any rash decisions.

Neogi agreed with the logic and came up with a plan to push the pig a bit farther forward. He had operators stack up oil in Thompson Pass and then let it go.

When they next measured, a dozen men participated. Their conclusion was unanimous: the pig had moved a dozen feet forward and was squarely in the receiver. Using a red paint marker that he had brought with him, Gibbs drew a little vertical hash mark on the pipe at the spot where the pig's magnets were centered. A measuring tape confirmed that from there, the pig was well clear of the ball valve. At 11:10, the techs opened the gate valve all the way. Neogi had expected to see the pig by now, but he had to wait a bit longer. It was time for lunch.

At noon, techs closed the valves on either end of the receiver, isolating the pig inside. They closed the bay door, activated the ventilation system, and began draining the trap. All told, at least fourteen men were in the building, more than twice the normal number. The Valdez crew was responsible for draining the trap, opening the door, removing the wax, and closing the door. The Baker Hughes and crane crews were responsible for lifting and moving the pig. Neogi and his integrity management crew were pretty much in the way, like a bunch of expectant fathers. But it was their project, so there they were.

At 12:25, the techs put on yellow Tyvek suits and respirators. At 12:30, one fastened a winch cable to the door. At 12:53, they approached the door. A minute later, the bolt above the yoke began turning. The yoke began to open.

The end of pipe went slowly from O to (O) to (O), such that there was room enough for them to pull the O out.

At 12:58, the tray door opened a bit and stopped. Two technicians shined flashlights into the trap and then kept opening it. It took four minutes to pull the tray all the way out, but it seemed like forever.

The first thing Neogi and everyone else noticed was that the pig wasn't covered in wax. It was "the cleanest a pig's ever been," he said later. Only five cups—not barrels, but *cups*—of wax were removed. The nearest drums got carried away empty. Techs, on their knees, wiped the pig down with rags.

Gibbs, meanwhile, examined the pig more closely. At first glance, he was relieved that the pig did not appear to have been hit by a freight train. It was intact. But as he looked closer, he discovered seven broken sensor heads. He called this typical, but still found it disappointing. At the butt of the pig, a blue light was blinking, and this was good news. The pig was still recording data. That all four odometer wheels were still present was also good news. Gibbs discovered no missing parts. A half hour later, though, he determined that the floating ring, an expensive two-part machined piece of aluminum to which the sensor arms are mounted, was bent, in the same way that a bicycle wheel may be tacoed. As a result, half of the pig's sensors were no longer aligned axially. Perplexed, Gibbs called it "a unique injury."

Once the pig was clean, the techs tightened ratcheting straps around it and connected them to the overhead crane. At 1:39, they raised the pig and put it down on a tray. Immediately, the Baker Hughes crew began retrieving the pig's data. The head of the crew, a cheery Colombian named Alejo Porras, plugged into it. Someone put a lid on an empty oil drum. Porras put his laptop on it and plugged in the other end of the six-foot cable. He began downloading 400 gigabytes of data. Then they waited. For the first time in days, the sun came out—but nobody noticed. The terminal crew was busy closing the trap door.

Neogi and his team headed back to town. Most took naps, and still looked zonked at dinner. Neogi took a hot shower, and finally admitted that he (a) had been awake at 5:00 a.m. and (b) was worn out. That night, food and beer brought the team back to life, but it was mostly pride that did it. Wasson was overjoyed that through the endeavor, Alyeska employees and contractors suffered no accidents or incidents (other than one speeding ticket), experienced no fender benders, required no Band-Aids. He credited good weather south of Atigun Pass. With beer, he managed to make it past midnight. Neogi—who forwent even a celebratory drink—was proud of the convergence that the long, difficult pig run had brought to his team. "All of us were focused on one thing, like a team sport," he said later. As a result, he felt melancholic. He was also proud of Alyeska's cooperation with Baker Hughes, and warmed by the way the industry got a feel for what one operator needed while the operator got the tool of its desire. Later that night, at a smoky bar, Super Dave Brown leaned back in his chair, removed his black baseball hat, wiped his head, put his hat back on, grabbed his bottle of Budweiser, and denied the rumor that he was retiring after this pig run. "I was sayin' I wouldn't do that again," he said, "but I'd do it again."

Porras finished downloading the data around two in the morning. He delivered it, on a black hard drive, to Matt Coghlan in his hotel room in Valdez. Then he returned to the terminal and began downloading a backup copy onto another drive. He fell asleep in his chair but woke up enough to confirm that the data were still moving.

The first thing the next morning, Dave Brown—who had over the last month put four thousand miles on his truck—drove back to Anchorage, headed home to Soldotna, and put up his feet for a couple of months. Porras finished downloading the backup copy at ten o'clock. The backup copy he kept in his backpack, and his backpack he didn't let out of his sight. The original data remained onboard the pig, but Porras took no chances.

That afternoon, as the weather began to clear, Ben Wasson—who had spent the last month getting up hours before sunrise and squeezing in catnaps when time allowed—flew out, and then headed to Hawaii for two weeks.

Before the Baker Hughes team left on Thursday morning, Devin Gibbs returned to the terminal and used his red paint marker to transform his hash mark on the pig receiver into a Canadian flag the size of a business card, thereby tagging the Alaska pipeline as he had tagged other pipelines on every continent but Australia. When Porras flew out, he told the attendants of the airport X-ray machine, unaware of how laborious it had been to gather the data on his drive, that it was "very sensitive." Later that day, the unnamed pig was crated, loaded in a truck, and driven back to Calgary.

On Thursday morning, Matt Coghlan finished his initial analysis and got word to Bhaskar Neogi that the tool had gathered data from 98 percent of the pipeline—confirming that it was indeed Alyeska's best pig run by far in recent history. "We stuck to our guns; we did the best job we've ever done," Neogi said. Alyeska's CEO sent congratulations. Neogi felt some relief, but he learned that on Thompson Pass, the pig actually bounced back a few feet when it hit the slack oil. He figured that this was probably how the sensor heads and floating ring broke. It made him, an admitted perfectionist, want to solve this problem, so that in 2016 he could attain a perfect run. Nevertheless—though it would be a month before he could declare the chance of a rerun beyond "remote," and a year before he'd be able to validate the pig data and conclusively say that it was a good run—he emailed his boss and preliminarily deemed the run "successful."

"That's the beauty of sports," he said. "You have a definite start time and a definite end time, and the real winner is clear. Pipe inspection doesn't work like that."

Neogi didn't leave until Thursday afternoon, and Hawaii was not in his future. He would spend the next weeks hiring a programmer, preparing for a PHMSA inspection, and heading to Calgary to check on the four-month analysis at Baker Hughes—which would determine where, over the next few years, Alyeska would have to dig up the pipe and carry out on the order of $40 million worth of work to prevent disasters like Enbridge, or BP, or Exxon, or San Bruno, or you name it, from occurring on TAPS. He would be managing the pipe's integrity, and his obligations would stifle a tropical vacation. At most, he'd find time to pig his fish tank. Until then, Bhaskar Neogi—the only guy in that deadleg town not hung over, or wired on coffee, or reeking of cigarette smoke; the only guy there with no inkling to fish or to ski—remained holed up at his hotel, eating bananas, listening to the grumble and beeping of loaders pushing snow around. He was waiting for the data.

10

BETWEEN SNAKE OIL AND ROLEXES

John Carmona, the proprietor of the Rust Store, once told me, "There's a subset of society that says, 'Hey, what I've got is pretty good, and I'm going to maintain it.' I think that attitude is what led me to open the store. I wanted to fix things up." He founded the Rust Store in January 2005. He was thirty. For a few months, Carmona ran the business from his home in Fitchburg, Wisconsin, storing products on two shelves in his garage, and packing them up for shipment on a tiny, folding card table. UPS picked up the packages from his front porch. Back then, he sold only four products: Boeshield Rust Free, Boeshield T-9, Evapo-Rust (in two sizes), and Sandflex Rust Eraser (in three grits). Business came so fast that he moved to an office in town. Within six months, he had twenty-five rust products in inventory—more than Home Depot currently stocks—and needed more space. The Rust Store moved into a building in Madison and, after a few years, moved to a larger building. Again, a few years passed, and business bulged. In the spring of 2012, Carmona relocated to a ten-thousand-square-foot warehouse in an industrial complex in Middleton, just outside Madison. From there, he now sells more than 250 rust products, tailored to tools, cars, boats, and so forth. He employs six staff, including his wife. The demand for rust products, he recently told me, is way steadier than the demand for football-themed products, which he also sells.

"In some ways, it's small and narrow, and in some others, it's broad and huge. It's in a thousand different places," he said. Carmona's voice is smooth, but he talks without finishing sentences, and then comes back and restarts, but not quickly. He just trails off and dutifully tries again. He's not hyper, but sort of spastic in a slow way. He seems either nervous or overwhelmed; likely the latter. "When the phone rings," he said, "you don't have any idea what the customer is gonna have a problem with—something tiny or the side of a building." Often, to diagnose a problem, he interrupts the customer, asking, Are you trying to remove the rust, or prevent it? Does the part need to be lubricated? Is it outdoors or inside? How large is it? Do you want a thin or thick coating? Does appearance matter?

"Rust is such a . . ." He trailed off. "There's not a lot of concrete answers." But he knows what he has, and he tries to help.

Carmona, who is tall, skinny, and goateed, studied business at the University of Wisconsin. Out of college, he did marketing for Gempler's, a vast Midwestern catalog of farm and agricultural supplies. When the internet boom came, he switched to e-commerce. Rust first captured his attention via two hobbies: fixing up cars and woodworking. Regarding the presence of rust in cars, he figured that in Wisconsin, rust was just part of the equation. (Here's guessing he worked on Fords.) Then the cast-iron tops of his jointer and table saw began to rust. "I thought, 'If there's two of my hobbies that deal with rust, there must be other applications that I don't even know about yet,'" he said. Neither of his parents was handy; his dad designed air-conditioning systems, and his mom was a homemaker. So Carmona investigated on his own.

I first called Carmona a week before the 2011 Super Bowl and asked if he'd received an email I'd sent with many rust-related questions. He said he had, and that there was a lot of stuff he hadn't attended to, as he was busy selling cheeseheads. He said it was the busiest he'd ever been in his life. For a second I wondered what cheeseheads had to do with rust. Even Home Depot doesn't sell cheeseheads.

Carmona explained that in addition to the Rust Store, he also runs a store called Wisconsin Goods, which sells, among other Wisconsin-themed items, cheeseheads ($18.50 for the original hat; $20 for a sombrero). "It's kind of a passion," he said. The Wisconsin Goods website, which lists eleven varieties of cheesehead hats, as well as cheesehead shirts, books, and rearview mirror ornaments, makes that clear. "It just blew up," Carmona went on. He described his cheesehead troubles, and his latest effort, on account of a national cheesehead shortage, to track down every last cheesehead from his suppliers' warehouses in time for the big game, to which the Green Bay Packers were headed. Because he could fit only a hundred cheeseheads in his car—"They are exc*eeeeeee*dingly bulky"—he'd rented a U-Haul and driven to Milwaukee. Now a blizzard—a record setter across the Midwest, bringing two feet of snow and winds that would render snowplows useless and ground a fifth of the nation's flights—was threatening his ability to keep moving.

He told me that the Rust Store is a much larger business, and that—lucky for him—things were quiet on that front right now, it being winter and all. During the winter, business at the Rust Store comes mainly from southeastern states. Until the rest of the country defrosts, Carmona stays busy by selling, in addition to cheeseheads: sharpening supplies, wool socks, muddlers (little wooden sticks for smashing fruit in cocktails), and beer hats, which are sombreros made from cardboard beer cases. When things warm up, Carmona sells antirust chemicals to grandmas with rusty cast-iron skillets and to drillers on Alaska's North Slope. It certainly is curious, the life of this frenetic Midwestern entrepreneur.

❧

Home Depot's rust expert is Cynthia Castillo. The handy daughter of a handy New Hampshire guy, she's been working at one Home Depot or another for twenty years. While going to college, she began working in the paint department of a store in San Diego. She became a department head, an assistant store manager, a store manager. She became a district manager, and opened a few stores in the Bay Area. She met her husband at Home Depot, in the building materials department. Now, at Home Depot's Atlanta mother ship, officially the Store Support Center, she works as the national paint merchant. To manage its tens of thousands of SKUs (stock-keeping units), Home Depot employs 150-odd merchants. Castillo oversees four hundred of those SKUs, including stains, paints, primers, waterproofers, solvents, cleaners, and rust converters and removers. She's been managing these products since 2009. I met her on a rainy day a few days before Christmas in Atlanta. She'd been up since five thirty in the morning and was cheery.

"I know about carpet, I know about tile, I know about draperies," she told me. "I mean, you just learn, because that's our business, home improvement." She knows about kitchens, too. And about America's rust habits.

The way Castillo sees it, the Mason-Dixon Line demarcates two approaches to rust. North of it, Americans paint, paint, paint. "They're constantly painting, so they're covering it up all the time," she said. South of it, Americans busy themselves removing rust and cleaning off resulting stains. "Rust is everywhere," she said. "It's in every city, every single town, every single home. There's some form of rust, whether people know it or not, behind the wall, outside." She also recognizes varying degrees of tolerance to it. Guys fixing up their motorcycles want it gone. Owners of rusty barns seem not to mind, and may even appreciate the look of rust. Castillo does not like the look of rust, especially not on the concrete around her backyard pool.

"A lot of people will use primers just to cover some of the rust up," she said. "Usually, if they don't remove it, it's gonna come back—but they still do it because they don't wanna clean it, and it's an extra step, so they use a primer. It's good for a short period of time. Some of our consumers may not take all the steps that they should take for a project, because they're in a hurry, so it's our job to explain that so they know the risks, that if you don't clean the rust off, Mrs. Smith, and you just paint a primer or a paint, it may come back." She sounded like she'd get along with Phil Rahrig.

Castillo said, "That's what our associates are trained to do. If they didn't tell 'em that, they'd be disappointed if six months down the road they look, and that stain that they covered up pops right through."

Home Depot sells three types of rust products: inhibitors, converters, and removers. Not all stores carry every product; Corroseal, for example, is available at stores in "marine clusters" such as Florida and the Great Lakes region, but not in Kansas. Nor does every store put the rust products in the same spots. Across the street from the mother ship at Store 121 (Home Depot operates nearly two thousand stores in the United States), I had Castillo show me—or try to show me—where they were.

Before going into any store, Castillo always dons an orange apron, like any sales associate. With her red fingernail polish, many rings, and bedazzled glasses, the orange apron went well. On the front, in white letters, it said, "Hi, I'm Cynthia. I put customers *first*." It looked festive. So did the store. With Christmas approaching, snowflakes, Snoopy, and reindeer hung from girders near the ceiling. Tinsel, wreaths, baskets of pine and berries, nutcrackers, plastic Santas, and snowmen for sale beckoned just past the entrance. Castillo walked past grills, poinsettias, a rack of orange five-gallon buckets, a promotional quarter-pallet of NFL-branded Atlanta Falcons Duck Tape (ten yards, $6.97). A dolly clacked and beeped as it went by. Over the PA system, a phone rang. The store was busy because on top of Christmas, painting season—otherwise known as spring—was around the corner.

Castillo went straight to aisle 47 (paint) to look for rust products. She stopped in front of a bay full of spray bottles, cans, bags, and jugs of chemicals designed for nuisances: grease, grime, graffiti, goo, gum, glue, crud, dirt, drips, sap, stains, spots, scuffs, "tough stuff," mildew, and residue. There was Krud Kutter (8 ounces, $1.49), Goo Gone (8 ounces, $3.48), Goof Off (4.5 ounces, $3.57), Dirtex (18 ounces, $3.98), Mötsenböcker's Lift Off (11 ounces, $4.99). Only four products mentioned rust, and they were all at crotch or knee height, in the land of wallpaper removers, and covered with dust. Castillo led me two aisles over, to a bay of adhesives between painter's tape and tubes of caulk. Below superglue, wood glue, contact cement, Liquid Nails, and regular old Elmer's glue, she pointed out Loctite Naval Jelly Rust Dissolver (16 ounces, $6.98) and Loctite Extend Rust Neutralizer (10 ounces, $5.67). They were at the same unpopular height, and easy to miss. When I told Castillo that they seemed to be in a funny spot, she seemed surprised.

Among merchants and VPs, Castillo has a reputation. At biweekly meetings, she always has new products to show off. This is not the case with the lumber department. The 2012 model two by four is the same as the 2002 version. As such, Castillo loses track of where her niche products end up. At least with me she did. To track down the other rust products in the store, she asked a sales associate in the electrical department—orange apron consulting orange apron—for help. "Do you have any rust removers?" she asked.

The associate led us to aisle 14 (hardware), where he stopped between tool belts and welding supplies, looked at a shelf of WD-40 spray, and said, "I guess not." But among the various lubricants were seven that claimed to prevent, protect against, penetrate, or eliminate rust and corrosion. Castillo had a sense that more lay hidden elsewhere. She walked to aisle 33 (sinks, showers, tubs, toilets), looking for cleaners. Like so many customers, she couldn't find what she was looking for. She asked another orange apron for help again and was directed to aisle 37 (kitchen appliances). There she found some cleaners, but none was tailored to rust. There was Steel Meister ("Put the stainless back in steel"), but that was it. There was also Glass Meister, a granite cleaner, a cooktop cleaner, and a washing machine cleaner, which seemed to beg a circular question.

I asked if rust products ever made it out from their dusty, forsaken shelves to the prominent positions of batteries, duct tape, and measuring tapes on the ends of the aisles. Castillo told me she'd never heard of it.

Before achieving prominent positions in the market, some rust products have been modified or removed entirely from the shelves—dusty or otherwise—by the US Federal Trade Commission. Regarding two motor-oil additives, the FTC told the products' makers to stop falsely claiming that the additives reduced corrosion in engines. The most famous case involved a $600 cathodic protection system for cars, sold by David McCready, of Pennsylvania. He called the system Rust Buster until *Dust* Buster sued him. Then he called it Rust Evader, Electro-Image, and Eco-Guard. It was two anodes hooked up to the car battery. In the early 1990s, corrosion engineers began denying McCready's claims that Rust Evader prevented or substantially reduced corrosion in cars. McCready responded with force. In response to a corrosion engineer at the University of Oklahoma who'd gone public, he sent a letter to the president of the university, who in turn shut up the professor. The scene repeated itself at Case Western.

At Texas Instruments, Bob Baboian began investigating, running a couple of experiments. In his lab, he scratched car doors and stuck the Rust Evader system on half of them. He put the doors on the roof of the building and on the beach in North Carolina. When he checked up on them later, he found no difference in the corrosion suffered by the doors with and without the system. GM and Ford did the same thing with complete automobiles at their proving grounds in Detroit. They found no benefit from Rust Evader. Baboian published his results.

"It didn't work," he recalled. "It was really a scam. An absolute scam. Just awful." The voltage in Rust Evader's anodes was so low that they protected only a couple of inches of steel rather than the whole car. "For this product to work, you'd have to string anodes all over the vehicle, two inches from each other," he said. Or live and drive underwater. McCready sent a letter to the president of Texas Instruments, alleging that Baboian didn't know what he was talking about. Texas Instruments stood behind Baboian.

In 1995, after looking at customer complaints and talking with Baboian, the FTC investigated. "They used to do demos in aquariums," explained Michael Milgrom, an FTC lawyer on the case. "I asked if someone had gotten it to work in cars." The FTC charged McCready with false advertising. When McCready denied the charges, the FTC prepared a case against him. The FTC had scientific evidence. Says Baboian: "It's not like we were just saying it. We had concrete data." The federal agency asked McCready what evidence he had.

McCready didn't respond to discovery requests. By April, fearing a civil penalty and/or fine, he relented and consented to a settlement. That summer, the FTC told McCready to stop claiming in any manner that Rust Evader was effective in preventing or substantially reducing corrosion in cars, and to stop making representations without true, competent, and reliable scientific evidence. To cap it all off, the FTC told McCready he was liable for $200,000 in consumer redress. The agency told him he had five days to come up with the money, and that if he wasn't able to sell his house in that time, it would take 20 percent now and give him six months for the rest.

Since then, McCready has moved to Arizona and back and taken up selling horological devices, also known as wristwatches. He and his wife sell them for $142.50 to ten times that under the brands Davosa and WestWatch. According to his websites, they're Swiss watches inspired by the American West and assembled in China. According to the same websites, McCready goes by d.freemont, has a nice mustache, and recently self-published a memoir in which he describes his "dislike of college purpose [sic] and posture" and his "rebellious nature towards narrow doctrine and limited points of view." The memoir does not mention Rust Evader.

When I finally tracked down McCready, I asked him why not. He said it wasn't an important part of his life. He said the FTC complaint was part of a conspiracy, that other products actually enhanced rust, and that politicians with great influence wanted McCready out of their hair. "I don't like being controlled by anybody," he said. Then he said he didn't even want to think about it, and asked that I not call him back. He said I'd need to pay him for his time.

Twenty-five minutes later, he called me back. "This is a painful part of my life," he said. "It's just like heartburn. I'm not interested in revisiting it. It's distasteful. There's a lot more than just the FTC and Rust Evader." Then he wished me luck with my book, said he had better things to do, and hung up.

Rust Evader is not dead, though. In Indonesia, it lives on under the name Neo Rust Evader. It comes with an eight-year guarantee. On a priceless company YouTube video that takes misrepresentation to the next level, the product is hailed as "US Technology." When I told Baboian about Neo Rust Evader, he said, "Now they can get away with it because cars are corrosion resistant. You put it on, and the person says, 'Wow, isn't this wonderful!' But if they don't have the product, the result's the same."

Back in Wisconsin, calls about rust still surprise John Carmona. Every week there's an oddball request, a question from a museum curator, or the head of maintenance at a golf course, or someone trying to get rid of rust stains at Lucas Oil Stadium, where the Indianapolis Colts play. Fielding a call from a guy with a driveway full of iron stains—"His wife's on him, and he's looking to patch his marriage up"—Carmona may joke around. Otherwise it's a pretty serious business.

"It's not like we get a lot of celebrities calling. Our customers are more middle income on down. People with millions of dollars don't typically have rusty things. Our customers tend to be pretty practical; they need help." They need help removing rust stains from their sidewalks. They need help restoring their tools, their cars, their houses. He recently answered a call from a guy restoring a Civil War canteen. "It literally looked like a big pile of rust," Carmona said. Indeed, a video of that restoration, on YouTube, attests to that.

In 2009, soon after Obama took office, Carmona got a handful of calls about preventing corrosion in guns. Rumor had it—at least in some places—that the feds would be knocking on doors, taking away firearms. So more than a few people decided to bury their guns in their backyards, and planned to keep them there until Obama left office. First, they wanted technical help, so they called the Rust Store. Carmona fielded the calls diplomatically. "We don't offer political advice," he told me, "but in this case, we'd still recommend not burying anything in the backyard." He thought a safe was the way to go. Still, he recommended an eleven-ounce aerosol can of Cortec CorrShield Extreme Outdoor Corrosion Inhibitor ($23.51) for the body of the gun, and a seven-ounce aerosol can of Lubricant & Rust Blocker ($10.99) for the lubricated parts, with the gun wrapped in a plastic bag before being placed in the dirt. (The Cortec stuff is also available in five-gallon containers, for those with really big guns, or a lot of them.)

"I just got an email from a lady who had a rust stain on a pink shirt—her favorite shirt. She was very skeptical, because she tried other products already, and we told her our products are guaranteed, and she reluctantly tried it, and a week later we got an email from her that said it worked great. She'd already bought another of the same shirt, so now she has two of her favorite shirt." This makes Carmona feel good. "We're not selling Rolexes," he explained. "Our products don't make Christmas lists. A gallon of rust remover isn't necessarily exciting." So he finds meaning in helping people. He's happy just to offer advice. "I would rather be honest than just get a sale that doesn't help anyone." On the other hand, he will ship anti-rust products anywhere in the country, by whatever means necessary. To small businesses, he's shipped orders that exceed a thousand pounds. He's taken thousands of calls. "Now when I look at a map, like Texas, I know of this city, or this town in Pennsylvania."

Carmona likes rust. He collects rusty things, buys rusty objects. He asks his buddies to bring in rusty stuff from their garages. They oblige, bequeathing old tractor parts, wondering what they'll look like without rust. He held a rusty tool contest on the store's website to raise the stakes. He keeps a stash. The stash is in two clear plastic tubs on a shelf in the warehouse. It contains a coffee can full of rusted nuts and bolts, pulleys, old saws, and rusted stainless steel, which he calls "a real treasure." He's on the lookout for a good piece of rusty chrome. "It's not like the fifties, when every car had chrome on it. Rusty chrome is not as easy to come by. I've got friends with motorcycles, but they don't want me testing on their motorcycles." He never goes to boneyards to look for rusty things. That's too easy, almost like cheating. Instead, he leaves tools in his driveway overnight—a hammer, say—then practices removing the rust from the hammer and removing the rust from his driveway. His staff practice too, with almost every product they sell. "Each time you remove rust from something, you gain a little bit more knowledge," he says.

For what it's worth, NACE recommends ammonia, hydrochloric acid, or oxalic acid for getting rid of red rust (Fe_2O_3, or hematite). For getting rid of black rust (Fe_3O_4, or magnetite), it recommends acetic or citric acid. But many variations will do.

"There's two groups of people that understand our business," Carmona said. "One says that's the greatest idea in the world, and another that doesn't deal that much with rust, and thinks, that's like the most bizarre idea I've ever heard of. You know, if you live in a city, in an apartment, you never deal with rust. If you live in the country, or on a farm, or work on cars, or live in the rust belt like I do, or on the coasts . . . Put it this way: we don't have a lot of customers from Arizona." About the business of rust, Carmona recognizes that a great portion of the country's rust problems are entirely preventable.

Carmona did not have a booth at Corrosion 2012 because he's not really in the corrosion industry the way the big players are. He's rust's Johnny Appleseed, a door-to-door rust salesman (via the internet). Aside from tool restorers, most of his customers are onetime only. He doesn't advertise much, because his customers find him when they need him. This suits him fine. In a world where few hardware stores have employees who know what they're talking about, where most read the copy on the package and consider that service, and no store carries hundreds of different rust products, his tiny piece of the market seems secure. "We have a lot of skeptical customers, and rightfully so," he said. "There's a lot of snake oil out there."

11

THE FUTURE

A t the International Time Capsule Society, at Oglethorpe University, in Atlanta, they're serious about the future. They keep a detailed registry of all known time capsules and encourage amateur time-capsule build-ers to sign it. They search for clues leading to the nine most-wanted time capsules, which, as fate would have it, have been lost to history. Most im-portantly, they advise time-capsule builders on archival storage techniques. They recommend a good, strong steel safe to create a cool, dry space in which to preserve artifacts. Get into details, and they'll recommend a product called Ageless Z100. Ageless Z100 is an oxygen absorber, used to create a hypoxic, or anoxic, environment for maximum preservative pow-ers. The product comes in sachets about the size of a wallet. The packets are inexpensive, because they consist of nothing more than iron filings in a permeable pouch, labeled NOT FOR HUMAN CONSUMPTION. The idea is that the iron will reliably, eagerly react with any oxygen in the time capsule, thereby keeping pesky oxygen from damaging any of the valuable artifacts also trapped within. Ageless Z100 thus becomes its own unintentional his-torical artifact. For in fifty, one hundred, five hundred, or a thousand years, when our descendants open our time capsules and find articles and docu-ments and relics revealing our culture, creations, and achievements, they will also discover those small white packets of iron oxide. Those little bags of rust. They'll know, almost before they know anything else about us, that oxygen was our enemy. That rust was our plague.

At MIT, they're even more serious about long-term preservation. When conceptual artist Trevor Paglen inquired about building the longest-lasting artifacts of human civilization—a hundred ultra-archival photos somehow stored in a time capsule mounted to a geosynchronous satellite—engineers at the NanoStructures Laboratory there decided to nano-etch the images into a silicon wafer and encase the wafer in five-inch discs of gold-plated aluminum. We know why they picked gold. Paglen's metagesture, which he calls *The Last Pictures*, includes many photos that bear witness to the force of nature: a dust storm over a Texas prairie, a waterspout in Florida, a typhoon in Japan, a Montana glacier, a guy surfing Mavericks. Among other things, it also features Leon Trotsky's brain, hela cells, the Great Wall of China, the Eiffel Tower, and at least two discernible mustaches. The capsule was launched aboard the EchoStar XVI satellite in November 2012. Paglen expects the thing to last until the year 4500002012—when the sun is expected to implode.

I asked John Scully, the UVA professor and editor of *Corrosion*, about advances in the field. He began with an analogy. "Did you know that in the 1850s there were fifty thousand deaths a year from exploding boilers? Then modern fracture mechanics came along." He said that corrosion has followed the same path. "A hundred years ago, corrosion was treated as an act of God. Now auto companies give you a ten-year warranty on paint. GM doesn't lose their shirt on that warranty, because they know damn well that it'll last in Michigan." He said, "The consumer can enjoy products that have safe lives. The engineer now has at his disposal bona fide winning strategies. Corrosion used to be a list of dos and dont's. An empirical listing, and now it's a scientific basis. It's almost equal to any other field. It's been transformed from act of God to empirical understanding, sort of like the medical field."

Yet according to a 2011 National Academies report called *Research Opportunities in Corrosion Science and Engineering*, to which Scully contributed, the field remains poorly defined and inadequately respected. Privately employed old-timers possess much of the field's institutional knowledge and pass it on slowly. Courses like those offered by NACE aren't far-reaching enough for mainstream engineers. As such, engineers tackling corrosion remain unaware of what they don't know. They overlook the field's various scientific journals, and certainly don't go to the technical research symposia at NACE or the Electrochemical Society. Corrosion is such an interdisciplinary field that most engineers fail to keep up with a fraction of research advances. According to Scully, many have horse blinders on. As a result of this gap, engineers keep making the same mistakes. "We're still in a response mode," he said.

There's an engineering joke about an old man who, as he aged, developed some serious health problems. He had trouble with his sphincter. To relieve him of his aches, doctors and engineers combined forces and devised an implant made of a fancy and expensive alloy. It worked and gave him ten more years of life. Two hundred years after he died, researchers exhumed his body. All they found was his asshole.

It's a joke about overdesigning. The world doesn't need $150,000 stainless steel or platinum-chromium assholes. It's a waste.

Gold-plated aluminum space discs aside, ask any engineer about the notion of the ultimate thing. If he's honest, he'll admit that it doesn't exist. Whenever we build something, we balance material properties (strength, weight, ductility) with human constraints (cost, durability, buildability, repairability). The science, or art, is finding the right balance. That's the job of an engineer. The responsibility of mature nonengineers is not to expect everything from every thing. If we accept imperfection and impermanence, upkeep becomes tolerable.

Yet we'd prefer to fastidiously tend to our upper lips more than our bridges and pipelines. Somehow, replacement—unaffordable as it is—remains more attractive than maintenance. Bhaskar Neogi thinks the conundrum has roots in American culture. He thinks Americans have ill-defined "macho." Macho, to him, isn't fighting power but thinking power. Mike Baach, the former corrosion industry executive who suggested a university corrosion program to Dan Dunmire, attributes the situation to recklessness. "We've been a reckless society," he said. "The more money you have, the more reckless you are." Everything degrades, he observed, but it's not just dollars at stake. "We can't afford it anymore. Is it a moral issue that our kids are gonna starve? Is it a moral issue when a busload of kids will get killed when a bridge falls? You can't separate them. It's moral, and it's economic." He pointed out that with exercise, you can get more physically fit, but you can't stop aging. "With corrosion," he said, "you can't get time back, but you can stop the clock."

More of our world is made of metal than ever before. The tab now stands at a record four hundred pounds of steel for every person on earth. As many sages have pointed out, the higher civilization rises, the farther it has to fall. Does our development suggest a type of insanity? Does our national approach, or lack thereof, to maintenance suggest laziness or even hubris?

When Robert Baboian wrote his graduate thesis on corrosion in the 1950s, corrosion was just a "tiny little" part of Rochester Polytechnic Institute's course work in electrochemistry. It was the same at all the technical institutes. For the last few generations, if you asked a mechanical engineer where you might learn about corrosion, he'd point you to a civil engineer. The civil engineer in turn would point you to a chemical engineer. The chemical engineer would direct you to a materials engineer, he'd direct you to an electrical engineer, and then *he'd* point you back to the mechanical engineer you started with. To most engineers, the future is not in corrosion. The future is in nanoengineering, genetic engineering, materials engineering, microbial engineering.

To Bernard Amadei, a longtime professor of civil engineering at the University of Colorado, this situation is pathetic. Amadei is a member of the National Academy of Engineering, the winner of both the Hoover Medal and the Heinz Award, and in 2012 was appointed as a science envoy by Secretary of State Hillary Clinton. The eminently practical son of bricklayers, he contends that engineering education in the United States is a failure because the aims of modern engineering are askew. He thinks that we're fixing problems we don't have and ignoring problems we do. He thinks that we now throw money at projects piecemeal, rather than design them right. A staunch believer in American ingenuity, Amadei nonetheless thinks we—humans—have a design defect. And he argues that the way we teach engineering contributes to the problem. "Here at Boulder," he said, "kids take Concrete 1, Concrete 2, Concrete 3, and then they go out, and they don't know how to mix concrete! They go, hmm, *pffff*!" On top of his thick French accent, he regularly pauses and says *pffff*. "It's deplorable! . . . We are virtual engineers. We're out of touch with reality.

"My colleagues still teach like it's the 1950s," he went on. "Traditional engineering is brute force. Dam that river. Dig that canal. If it doesn't work, try harder. . . . Civil engineers have to build something big: my tower is bigger than your tower." Considering the condition of American infrastructure, which earned a D from the American Society of Civil Engineers, he calls this approach a "technical wasteland." As evidence, Amadei cited recent train travel. He took the Acela on an hour-and-a-half trip, and was an hour and a half late. He wrote a letter to Amtrak and got his money refunded.

To get engineering students back in touch with reality and to give them a conscience, Amadei founded Engineers Without Borders. The organization now has twelve thousand members engaged in more than four hundred projects—mostly water or sanitation related—in forty-five countries. On the Crow reservation in Montana, he started a company that makes cinder blocks out of local clay. In Afghanistan, he started a company that makes fuel out of waste paper. In Israel, he showed Bedouins how to use renewable energy to make cheese.

Amadei has been rethinking engineering schools, too. He told me that the best engineering program he's ever seen isn't where he teaches at the University of Colorado, or Stanford, or MIT, but at KIT: the Kigali Institute of Science and Technology, in Rwanda. It was started in 2004. All engineers there begin by spending three months in a village. When they come to school, they're asked what they can do to solve problems. They do the same thing over the next three summers. Then, to get their degrees, they have to demonstrate what they've done to improve that community.

This kind of approach makes Amadei's eyes open wide. "I see great opportunity," he said. "We need a new mind-set for a new ballgame with new players." He's eager to get busy "reengineering engineering." To do so, he'd begin by insisting that engineers take a wider range of courses, because he thinks that engineering education in America is hampered by its narrow base, churning out only specialists. He's in favor of fewer universities and more vocational schools. He'd also support any efforts to educate more engineers, who constitute only 0.5 percent of the American population, especially female engineers.

John Scully, Luz Marina Calle (the head of NASA's corrosion branch), Paul Virmani (the author of the Department of Transportation's 2001 cost-of-corrosion study), and Dan Dunmire all see the situation similarly. In that 2011 National Academies report, to which they all contributed, they identified corrosion education as a major national concern. In a typical engineering program, "students' exposure to corrosion issues is limited to a single lecture in corrosion in an introductory materials science class," they wrote. "In many cases, lectures on corrosion are not offered because of time constraints and the demands of other topics. At many universities, even students majoring in materials science and engineering (MSE), who should be trained in materials selection, receive limited exposure to the topic of corrosion because only a fraction of MSE departments have even a single course on corrosion in their curriculum." They blamed overcrowded curricula, scarcity of qualified faculty, and a lack of awareness. They praised NACE for offering corrosion modules at educational summer camps for high schoolers, and pointed to the Corrosion And Reliability Engineering (CARE) program at the University of Akron, the country's only undergraduate corrosion engineering program.

Because of corrosion, engineers are rethinking the materials that have gone unchanged in structures for a long time. One such new material is extruded structural composite, or thermoplastic lumber, as it's usually called. It's been around for twenty-five years, mostly in nonstructural arenas: picnic tables, park benches, decks, boardwalks. New York City built a four-hundred-foot pier out of the stuff in 1995, but a year later, it was hit by lightning and burned down. Nevertheless, the new material has been touted as superior to wood because it doesn't splinter, crumble, crack, warp, or rot, and doesn't require the chemical treatment—chromated copper arsenate and pentachlorophenol—that pressure-treated wood does. In fact, it's impervious to infestation by insects, it doesn't mushroom like wood when pounded, and in stress tests, it outperforms oak after exposure. Plus it's recycled.

In 1998 Axion International, a small company in the heart of New Jersey steel country, showed off what the material could do on a twenty-four-foot bridge in Missouri. The bridge had steel I beams supporting a wooden deck, which was replaced with Axion's thermoplastic boards. They cost twice as much as new wooden planks. By 2006, though, the plastic seemed a bargain, because it required no maintenance and was still in perfect shape. The market, though, was not impressed. Where the three years leading up to the bridge saw Axion stock soar from $150 a share to over $2,000 a share, the next three years saw shares fall to below $13.

Not taking the hint, Axion went on competing with wood by manufacturing railroad ties. The company tested them at the Federal Railroad Administration's Transportation Technology Center, in Pueblo, Colorado. Since 1999, the center has run 1.5 billion tons over them, in 39-ton cycles. The ties have yet to break or warp. They resist plate wear, hold spikes, and maintain their gauge. Armed with such evidence, Axion has sold 200,000 such ties (a mere 65 miles of track) in Dallas, Kansas City, Jacksonville, and Chicago—where salt rains down every winter courtesy of that God-like agency called the DOT. Because the plastic ties last at least four times as long as wooden ties, at only twice the initial cost, it seems a sure bet that Axion stands to gain from an industry that replaces 20 million of them each year.

In the last decade, though, Axion's engineers, working at Rutgers University's materials science lab, developed a new polymer three times tougher than the old mixture. By blending HDPE plastic (#2) with old car bumpers and 11 percent fiberglass, they created a material stronger and more flexible than steel, but only one-eighth as dense. It floats. Its light weight makes it cheaper to transport than steel. Most importantly, it doesn't rust.

Rather than compete with humble wood, Axion sought, ambitiously, to compete with steel. It began, in 2002, with an elegant fifty-six-foot arch bridge in New Jersey that could support thirty-six tons. Even then, though, the structure's strength came from laminated boards bowed into a van Gogh arc. Not until 2009 did Axion demonstrate that its thermoplastic could be fashioned into, and employed as, regular old-fashioned I beams. But this time, it did so dramatically. At Fort Bragg, North Carolina, Axion's plastic I beams were used in a forty-foot bridge spanning a muddy creek. The US Army Corps of Engineers drove a dump truck full of rocks over it. Then the army drove a seventy-ton Abrams M1 tank onto the bridge. Halfway across, the tank stopped and parked. After testing—the bridge was wired up with a couple hundred sensors— engineers determined that it could safely handle one hundred tons. Dan Dunmire, the Pentagon's rust ambassador, was behind that trial. The next year, designers in Virginia put Axion's plastic I beams to use in two plastic railroad bridges, the world's first. They're forty feet and eighty feet long, and look as good as truss bridges out west. Axion's greatest coups, though, took place in 2011. In York, Maine, a tiny fifteen-foot bridge became the first plastic bridge on a public US highway. And in Scotland, a ninety-foot bridge, Axion's best looking yet, erected in only four days, became Europe's first recycled plastic bridge.

The North Carolina bridge cost just under $500,000, which, in bridge jargon, was $675 per square foot, or a pretty good deal. Axion says that the thermoplastic bridge cost half as much as a comparable steel bridge, and will last more than twice as long, with no hassle. It says that its thermoplastic is resistant to acids, salts, abrasion, and even ultraviolet light, which destroys only 0.003 inches of the plastic per year, far less than corrosion would destroy on a steel bridge. According to Dunmire, this is the bridge of the future. It's got a lifespan of at least fifty years, and requires no maintenance, because it doesn't corrode.

Ironically enough, Axion manufactures the plastic in Portland, Pennsylvania, on the Delaware River, thirty miles northeast of Bethlehem, where steel for the Golden Gate Bridge was made eighty years ago. Axion recognizes the modern opportunity before it by pointing out that the global demand for infrastructure upgrades is in the trillions of dollars. As a result, it now calls its patented thermoplastic technology "disruptive," which is part of the reason that Dunmire likes it so much. The material will disrupt the American steel industry, and maybe the tasks of bridge inspectors. (It remains to be seen if it will disrupt the endocrine systems of the biological systems nearby.) But mostly, it disrupts the constant battle against corrosion.

Plastic won't work for engines of automobiles, but it will for their exteriors and maybe their frames, making them lighter and more efficient. Boats, as mortal as humans, remain destined to be made of metal. Same for pipelines and cans. Alternatives remain a long way away but are not unimaginable. Most statues, the most durable of our creations but also the most old-fashioned, seem fated to be fashioned of metal forever. Calder or not, they'll go on corroding. Like babies or old men, they'll always need some form of care: cleaning, waxing, blasting, or artificial patination.

~~~~

To deal with corrosion today, the authors of the 2011 National Academies report made some suggestions. As they saw it, of the various government agencies dealing with corrosion, only the DOD and NASA had comprehensive, well-funded plans. The DOD program, they wrote, "might serve as a model for what should be sought in other large government organizations." Each federal agency or department, they continued, should draw up a corrosion road map, addressing the four "Corrosion Grand Challenges." These were: (1) developing corrosion-resistant materials and coatings; (2) predicting corrosion; (3) modeling corrosion through lab tests; and (4) outlining a corrosion prognosis (in other words, when objects will need repairs, overhaul, or replacement). This they called a "national corrosion strategy." They said an "overall federal effort," including support from the Office of Science and Technology Policy, should address the various road maps. They called for documenting current federal expenditures on corrosion research and mitigation, and then funding a "multi-agency effort for high-risk, high-reward research." They called for collaboration among departments and agencies with state governments, professional societies, industry groups, and standards-making bodies. They called for the creation of a corrosion consortium much like the Human Genome Project, to store the vast amount of thermochemical data on metals and the ways they rust in various environments. They hinted at the National Institute of Standards and Technology's success in a similar endeavor, and pointed out that since corrosion research tended to find its way into applications at a "glacial pace," this would revolutionize the field. They agreed with the Defense Science Board Task Force's conclusion that "an ounce of prevention is worth a pound of cure." They wrote, "The committee believes that government-wide as well as society- and industry-wide recognition of the scope of the corrosion problem and a well-defined, coordinated, and reliably resourced program would have a high payoff for the nation."

Nobody, the authors wrote, seriously understood the need to conserve national resources. "Corrosion affects all aspects of society, in particular, the areas where the federal government is investing: education, infrastructure, health, public safety, energy, the environment, and national security."

My own opinion, after a couple of years pondering rust, is that the Department of Transportation ought to turn to galvanizing bridges in the great many places where possible and where paint isn't cutting it.

The Food and Drug Administration ought to insist on part-per-trillion analyses of the endocrine-disrupting agents at the heart of corrosion-preventing epoxies lining food and beverage cans. In the meantime, it ought to put labels on all cans stating that pregnant women should not consume foods or beverages from cans.

Congress ought to tighten minimum pipeline inspection standards and grant more enforcement powers to the Pipeline and Hazardous Materials Safety Administration. Opposing the construction of new pipelines is silly. Oil is only getting more valuable and likelier than ever to be shipped out of the ground to hungry consumers—myself included—by one means or another. Pipelines are the safest way to deliver the oil. Demanding that we know the condition that pipelines are in, on the other hand, is not silly. It's what Senator Warren Magnuson wanted to do when he said, "Keep the Big Boys honest." To keep 'em honest, we ought to demand more frequent inline inspections (for leaks and metal loss), lower metal-loss intervention criteria, make the information public, and impose higher penalties for failing to abide. Fines need to be proportional to the dollars flowing out of the pipe; otherwise they're just chump change. Offshore drilling ought to be contingent on stricter regulation. It is, after all, our oil, leased from our land, flowing through tubes on long stretches of yet more of our land. Following our rules is the least the industry should do. If Alyeska can send a smart pig from Prudhoe Bay to Valdez every three years and in so doing keep the Trans-Alaska Pipeline from suffering a corrosion-induced leak, every other pipeline operator ought to be able to do as much.

The president ought to direct more funds to Dan Dunmire's Pentagon office, considering the returns we'll get. He ought to support the plan by the National Academies, calling for a civilian version of Dunmire's office. He ought to say the word *rust*.

Like any environmental story, dealing with rust should give us a little more respect for what's public, a little more regard for the future. It should also educate us, showing that the *right stuff* is not thinking with your gut but the result of engineering-like analysis. Admiring only the shiny and new is what spoiled babies do. Admiring the practical and effective takes maturity. While children admire Buzz Lightyear for his bravery and strength and improvisation, the rest of us can admire Robert Baboian, Bhaskar Neogi, and Ed Laperle. Don't we need some engineering heroes? Finally, unlike so many bleak environmental stories of the moral and practical variety, we may see results long before we degrade and die.

EPILOGUE

Before they got caught and set off the most expensive, most public, and most symbolic rust battle in American history, Ed Drummond and Stephen Rutherford spent that night in May 1980 sharing one sleeping bag and one down jacket high over the historic entrance to America. Shivering, they distracted themselves by reading Emerson, Dickinson, Frost, and Angelou out loud. Before dawn the next morning, a police officer leaned a twelve-foot ladder against the Lady, climbed to the top, and put a finger to his lips. "Psst," he said. "*Psst!* Are you cold?" Drummond and Rutherford were suspicious; the other cops had spent most of the night taunting them. But they saw sincerity in this man. The cop's name was Willie. He offered the climbers wool blankets and hot coffee. Drummond rappelled down to him, retrieved the offerings, and never saw the man again. He called him an angel in disguise.

In court the next day, having charged Drummond and Rutherford with "malicious damage to government property," the judge asked, "Who will pay this bond?" Just then, the door swung open, and in came Bill Kunstler, the most hated lawyer in America. "I will," he said in that deep scratchy voice. His entrance was perfect. Dramatically, he posted the deed to his house on Gay Street. The judge put his head in his hands and said, "Oh

God, it's you." He had the case moved far from his jurisdiction. To Drummond, Kunstler said later, "This is how things change, this is how things get done."

As a result of the publicity surrounding the stunt, authorities in California moved Geronimo Pratt from San Quentin to San Luis Obispo, and prison guards there gave him hell. Drummond visited him a few times and saw him deteriorate. Pratt put on weight, became cynical. To focus on the future, Drummond promised to take Pratt up El Capitan if and when he was released.

Back in the Bay Area, Drummond and Rutherford fulfilled their community service by taking a handful of kids from Oakland into the Sierras, to a granite cliff called Lover's Leap. There, they spent a weekend teaching the kids how to rock climb. On December 8, 1982, two years after John Lennon was murdered, Drummond climbed one of the Embarcadero Center towers and hung a banner that said Imagine No Arms. After San Francisco authorities promised to leave his banner up, he descended. Then they tore it down. A year later, he climbed the tower again to hang a huge button that said Yes on 12. It was in support of Proposition 12, which would freeze nuclear weapons development. While Drummond was climbing, the building manager threatened to cut his rope. While he was getting arrested, a journalist yelled out, "What's your next stunt, Ed?" Drummond realized he was done climbing buildings for political purposes.

For work, though, Drummond continued to climb buildings. He spent years working as a steeplejack in San Francisco. He got married on the roof of Grace Cathedral. When he learned of the extensive plans for the restoration of the Statue of Liberty, he thought about bidding for the job. "We'd have done it for a million," he said.

In 1997, twenty-seven years after he was first jailed, Pratt was exonerated for the murder of Caroline Olsen and awarded $4.5 million for false imprisonment. Drummond was right: he had been framed. Having spent nearly half his life in prison, Pratt wasn't about to waste a few days on a granite monolith. In Los Angeles, when he was released, the crowd was so big and boisterous that Drummond only had a moment to say to Pratt, "Remember, we have a climb to do." Pratt, carried by the crowd, replied only, "Oh yes, climbing." Then he was gone. Drummond wrote Pratt a couple of letters and never heard back from him again.

Pratt died in the summer of 2011. When he did, I was heartbroken to inform Drummond of the news.

Since that cold night, Drummond hasn't been back to the spiffed-up Statue of Liberty. He lives in San Francisco. Rutherford still lives in Berkeley. After thirty-four years of teaching science (including corrosion), he recently retired. He still has the suction cups they used. The two have met a few times since their Statue of Liberty days.

Less than a month after Superintendent David Moffitt caught Drummond and Rutherford climbing, Croatian nationalists detonated two sticks of dynamite inside the base of the statue. Moffitt dealt with that surprise too and stayed at Liberty Island through the long restoration. He then spent two quiet years as superintendent of Colonial National Historical Park, near Jamestown, Virginia (out of the bombing and VIP zones), and four years in Washington as assistant director of visitor services for the National Park Service. He retired, moved to Williamsburg, Virginia, and started a landscaping business called Jusdavid Landscaping. He was the only employee. No climbers ever interrupted him. He retired a second time in 2003. Not long ago, he wrote me, "After all these years and with the exoneration of Geronimo Pratt, that protest, albeit illegal, seems almost admirable." At seventy-four, his knees have given out, but he still likes to get down and pull weeds. He just needs help getting back up.

Robert Baboian, the corrosion consultant, has been back to the statue many times to check up on her. He was there in October 1986, when NACE dedicated its National Corrosion Restoration Site plaque, and he was there a year later, only to discover that the plaque was corroding. It was turning green. The American Society for Metals and the American Society of Civil Engineers had installed their own historic landmark plaques in 1986, and they were doing fine. When ASM and ASCE executives heard about NACE's rusty plaque, they laughed their butts off. Baboian took care of the problem. He had a New York sculpture artist strip off the old interior-grade coating and apply a durable, exterior-grade coating to it. The plaque has survived okay since. It's not shiny and polished like ASM's, but a dull, muted brown. It's not corroding, though.

Neither is Lady Liberty. For the first ten years after the restoration, Baboian inspected her annually. Every time, like some dignitary, he went straight to her torch. He poked around her crown. Then he descended stairways to her toes. From the ground, he looked up with a telephoto lens. After the mid-1990s, he resorted to biannual inspections. They take two days. Today, with the torch closed to the public, he revels in going up there. He's seventy-nine years old and still vibrant. The statue is 128 years old and comparable. "She's doing good, really good after twenty-five years," Baboian said. "The stainless steel is holding up well. It doesn't leak on the inside anymore." Over the years, he said, most of the baking-soda-induced stains have faded, but a couple of prominent ones remain visible on the Lady's neck and right temple. Guano accumulates in the numbers on the statue's tablet but washes away with rain.

Baboian said something that surprised me. "Pretty soon, the gold leaf will have to be redone." The gilded torch has worn thin and, according to Baboian, will need fixing by 2016. Baboian had wanted the torch to be electroplated with nickel and then gold, like the statue atop the dome of the Rhode Island State House, but the restoration committee had resisted. When the torch gets re-re-re-re-done, Baboian hopes that someone heeds his advice.

Officially, Baboian retired years ago, but from his home in Rhode Island he still consults under the name RB Corrosion Service. He takes on projects he likes, preserving an iron basilica in the Philippines, a bronze Buddha in Japan, the USS *Monitor*, the Enola Gay, the Capitol dome. He tells a lot of kids that they ought to be corrosion engineers. "You won't ever have a problem getting a job," he says. "Never ever. And you'll get paid well." He said his career in corrosion has treated him well. He owns two boats. He regularly gets out fishing off Cape Cod and often makes it down to Marco Island, Florida. In the most defiant act of all, he roots for the Red Sox.

Three years after Bertha Krupp's 150-foot stainless steel schooner was seized by Britain at the start of World War I, she was sold at auction. The Norwegian buyer gave *Germania* to his brother, who renamed her *Exen* and sailed her to New York. As any boat owner could have predicted, he went bankrupt. In 1921 his estate sold *Exen* to Gordon Woodbury, the former assistant secretary of the navy. He renamed her *Half Moon*, after Hudson's ship, and refitted her lavishly. Off of Cape Charles, Virginia, the next year, en route to the South Seas, Woodbury sailed through such a storm that he nearly drowned. His quartermaster was washed overboard and lost. Woodbury limped home under tow from a Standard Oil tanker and sold the boat promptly. The next owner cut off the lead keel and sold the hull for scrap. The buyers renamed her *Germania*, towed her to Florida, and employed her as a floating restaurant and dance hall on the Miami River. Damaged in the hurricane of 1926, she sank. Raised, she was bought by a Captain Ernest Smiley and renamed *Half Moon* again. Smiley anchored *Half Moon* on a reef miles off of Florida, lived on board with his wife and son, and employed the vessel as a Prohibition-era cabaret. In a severe 1930 storm, the Smileys abandoned ship, and *Half Moon* broke free and ran aground less than a mile off of Key Biscayne. The Smileys were rescued, but *Half Moon* was not. For generations, she buried into the sand and was forgotten. Modern divers speculated about her identity, thinking her, perhaps, *Haroldine*.

Mike Beach, an energetic scuba instructor and guide earning a master's degree from the University of Florida's Rosentiel School of Marine and Atmospheric Science, began writing his thesis about *Half Moon* not long after she was identified. Fifteen years and at least a hundred visits later, he took me out to see her. From a sandy park covered with palm trees, we hopped in sea kayaks and paddled out. The morning had been clear and calm, and the tide was slack, but the wind picked up halfway out and slowed our progress. Swells picked up, and the water, until then clear, got murkier. Miami was visible behind the barrier island. Over the monotonous paddling, Beach told me he'd heard that hammerheads and bull sharks were out in the area. "We'll be lucky if we see sharks," he said. "If I see a shark, I'm jumping in." Then he showed me scars on his left leg, and said he used to be a handsome guy. After a shark attack in 1996, he got four hundred stitches in his face and leg. "Lightning never strikes twice," he said.

When we tied our kayaks to the buoys marking the wreck twenty minutes later, I was not as excited to go in. I put on fins and goggles and a snorkel. Then I followed—or tried to follow—Beach. We started at the stern, which was maybe ten feet below the surface. He dived down into the dark blue water, pointed to the tip of the ship, and vanished beneath her skeletal frame. I kicked to keep from rising. She didn't look stainless. She looked greenish and barnacle ridden. Then again, stainless steel doesn't do well in saltwater. I popped back up on the surface, waiting for Beach. The man has lungs. The current pushed us toward the bow. Down he went again, and I followed. Back on the surface, when he mentioned a stingray and poisonous coral, I lost interest in seeing the bottommost hull of *Half Moon*. Five feet down, I could see the structure of her frame, the length of her stringers, the width of her beam. For fifteen minutes, I admired her form, contemplated the long journey that ended here and was still ending.

Late that night, I found out that Beach—professor, triathlete, licensed coast guard captain, PhD in maritime history—is also a good drinker. Halfway through a bottle of Laphroaig, he told me that those were "pretty sharky conditions." All for rust, I joked.

Seven months later, the British Stainless Steel Association celebrated the one hundredth anniversary of Harry Brearley's discovery. In Sheffield, on the side of a building, an artist painted a four-story-tall mural of the rebel whose persistence changed our expectation of metal. Industry magnates and a Member of Parliament spoke. An exhibit on Brearley opened at a museum. At a lavish dinner, his grand-niece, Anne Brearley, spoke briefly to great applause. The next day, at the University of Sheffield's Advanced Manufacturing Research Centre, his great-grandson, Warren Brearley, unveiled a plaque dedicating the new Brearley Suite.

<hr />

Forty-one days after the pig arrived in Valdez, Alyeska pumped 600,000 barrels of oil through the Trans-Alaska Pipeline in one day. It hasn't pushed as much crude through since. In the summer of 2013, when daily flow rates regularly dropped below 500,000 barrels, and often under 400,000 barrels, Bhaskar Neogi got the first hint of his data. Baker Hughes's preliminary report said that TAPS had no imminent threats. Investigations along the pipeline confirmed the pig's accuracy.

In early September, a 10-inch steel disc showed up in Valdez, perplexing many. Alyeska figured out that it was part of a valve at mile 385 that had been used, in the 1970s, to demonstrate the integrity of the pipeline (using water pressure). In 2012 Alyeska had encapsulated the old valve— apparently just in the nick of time. When the threaded O-ring failed, and the steel disc began swimming south in a stream of crude, no oil was spilled. Examining pig data, Neogi reassured higher-ups that 150 other such valves along the pipeline were in fine shape.

Later that month, Neogi was promoted to Alyeska's senior director of risk and compliance. The move put him among the nine senior company leaders who report directly to the president. Roughly a hundred people work for him, and he travels as much as ever. As of early 2014, though plans were afoot for a 2015 pig run, Alyeska still hadn't hired a replacement as the pipeline's integrity manager.

Neogi's fish are doing fine.

Ben Wasson also landed in Alyeska's Regulatory Affairs department. Todd Church, the manager of the Operations Control Center, who said that there was enough oil at Prudhoe Bay to get him through his career at Alyeska, moved on to another company before finding out.

At Baker Hughes, Devin Gibbs has been so busy inspecting pipelines that when I called his office and said I was trying to reach Devin Gibbs, the receptionist laughed and said, "And how's that going?" Alejo Porras has been just as busy, inspecting pipelines from Ontario to Colombia.

Bill Flanders and Dennis Hinnah, PHMSA's senior regulators in Alaska, who saw their roles as "trust but verify," retired within a month of each other.

Since Neogi began planning the 2013 pig run, California, as serious about its water as Alaska is about its oil, has pioneered the use of a modified MFL pig in its mortar-lined water mains. San Francisco did it first, pigging thirteen miles of three parallel fifty-inch pipelines running across the San Joaquin Valley in 2010. Through these pipes, built as early as the 1930s, travels water from Hetch Hetchy to millions of people. San Francisco's public utilities commission hired a company called Emtek (since acquired by Pure Technologies) to develop the pig, and Emtek came up with a custom, retractable MFL pig that looks something like a giant Tinker Toy made from tent poles. Thanks to fiercely strong magnets and modified algorithms, the pig can detect anomalies in steel covered in up to an inch of mortar, as these water mains are. Since there's no pig launcher or receiver in the San Joaquin pipelines, the logistics get ugly. To run the pig through a line, the line must be drained, dewatered, and opened. Through a manhole, two thousand pounds of parts are lowered into the pipe, and there assembled into a pig. A custom electric all-terrain vehicle, with big rubber tires, is also lowered into the pipe—and this pulls the pig. It's not fast. When the driver of the ATV gets to a butterfly valve, through which neither tool nor ATV can proceed, both are disassembled, passed through it, and reassembled on the other side. This takes two days. As such, in a monthlong shutdown, San Francisco may get only a week's worth of pigging and collect data from only five miles of pipe. But San Francisco has found it valuable. Thanks to the pig, the city found more than one thousand anomalies in the encased steel. Most were under 30 percent wall thickness, but ten were around 40 percent, and one measured 90 percent. San Francisco found the pig so valuable that it bought the tool, put it in a warehouse, and has plans to keep using it until all three lines have been inspected.

In an unusual show of Californian city sisterhood, San Francisco loaned the tool to San Diego in late 2011, so that the county water authority could inspect a six-foot water main in San Marcos built in 1958. In fifty years, it had been inspected only a handful of times, and then only visually. The status of exterior corrosion on the pipe was a mystery. Up steep sections of this pipe, the ATV and pig were winched. As expected, the logistics were nightmarish. In a week, the tool covered a mere five miles—which made the sucker in Alaska seem to be moving at warp speed. But it found severe anomalies, and left employees of the San Diego County Water Authority satisfied and impressed. They had plans to use the tool again, on a different section of their system, in 2014.

Myron Shenkiryk, Pure Technologies's western regional manager, told me that since the California utilities began pigging some water mains, a dozen other water utilities, from all over the country, have gotten in touch. Apparently the company has a monopoly; the big players in the pigging industry don't like the logistical nightmare involved with pigging water lines. Water money also pales compared to oil and gas money. Securing that monopoly, Pure Technologies could inspect sewage lines, which, on the plus side, usually have pig launchers and receivers. "Most agencies don't even know that the technology exists," Shenkiryk said.

⌒

Since the spring of 2011, Ball has spat out more than a hundred billion cans. The company did not invite me back to Can School, which remains popular, and for which Ed Laperle still leads tours of the corrosion lab. Laperle has been as busy as ever—testing eight hundred beverages in 2013. "Everybody's looking for the next Rockstar," he said. With at least a couple of the experimental beverages, he encountered what he called "unique corrosion scenarios," but insisted that the details were confidential.

Alysssha Eve Csük spent the year away from the Bethlehem Steel works, teaching photography at Appalachian State University. She moved out of her studio and bought a house just outside town. She was selected as a featured artist by Hahnemühle. (And contracted for this book's cover image.) As always, she yearns to get out and shoot more.

Cynthia Castillo, Home Depot's buyer of rust products, transferred from the paint department to the decor department a month after I saw her. The new rust product merchant came from the electrical department.

For the first time in years, John Carmona of the Rust Store hasn't upgraded to a bigger facility. He has, though, hired two more employees and now stocks nearly three thousand SKUs. Business is so good that he has little time to test the many samples piling up on his desk. He's had no cheesehead emergencies.

Phil Rahrig, pushing the agenda of the American Galvanizers Association, has been playing as much defense as offense. Since April 2013, the defense has regarded nearly one hundred galvanized rods in the east span of the new San Francisco–Oakland Bay Bridge that failed prematurely, and which led many observers to jump to conclusions about not just the few thousand other rods but also the fundamental worth of galvanizing. Annoyed by the innuendo, he called the failures a manufacturing error. Apparently the company that manufactured the rods lost some of its records and accidentally heat treated the rods twice, weakening them. The galvanizer, he insisted, was innocent; the process, he said, still as valuable as ever. Rahrig's offense has entailed taking on the stainless steel industry in addition to the paint industry. In New York, that's meant convincing authorities that galvanized parts were suited for the Tappan Zee Bridge. "Stainless is not Kryptonite," he said, as if I were considering purchasing a few hundred tons of the stuff. Then he added, "That they were looking at stainless tells me they were considering maintenance costs of the project. That's very good news." Indeed, Rahrig suspects that government agencies have begun to more seriously consider life-cycle costs. The evidence: in the last decade, demand for galvanized parts has increased 60 percent, to four million tons annually. Further evidence: Rahrig's loathing for the DOT has ebbed a bit. A great many of those tons have gone toward the solar industry, building frames for so many panels. A lesser amount has gone toward Denver's new light-rail system, whose stations are painted beige. They're not just painted. They're duplexed. Said Rahrig: "I know the galvanizing is underneath."

NACE got a new president and began reforming the disbanded committee on National Corrosion Restoration Sites. The organization is considering commemorating a battleship as well as the restored dome of the US Capitol. Last I checked, it had not yet considered the material for its plaques.

Maren Leed, who got corrosion in front of Senator Daniel Akaka and into federal code, works as a senior advisor at the Center for Strategic and International Studies in Washington, DC. "It's not like I'm a corrosion hero," she said. "I was just doing my job." She credits Congress. "Congress is a pretty screwed-up place, but every now and then they can do some okay things." Cliff Johnson, the public affairs guy from NACE who brought corrosion to Leed's plate, now works as the president of Pipeline Research Council International—where he looks at pigging. Leed called him the luckiest non-lobbiest she's ever met. Had the new ranking member of the Senate Armed Services Committee been from, say, Wyoming, the Pentagon's corrosion office never would have been created. But it was a good idea, and the conditions were right. As far as congressional sausage goes, this was pretty clean.

Dan Dunmire, who as a result of their efforts has been running with corrosion for the last decade, has since impressed Leed. "I was worried that because of him being an odd duck, it would hamper his ability to get things done in the department," she said. "But he's got an infectious passion, and he's eternally optimistic." Larry Lee, Dunmire's chief of staff, said, "It surprises a lot of people that we were able to pull off a project as mundane as this." Then he added, "We created a monster!"

When the rust exhibit opened at the Orlando Science Center in March 2013, Dunmire couldn't go because of federal sequestration. He'd intended to speak at the ribbon-cutting ceremony. Instead, he flew to Orlando on his own funds and kept a low profile. "Nothing can stop me," he said. "Nothing." He learned that Saudi Aramco, Saudi Arabia's immensely valuable national petroleum company, wanted to support one such exhibit in Saudi Arabia, and that NACE yearned to get behind one in Houston. Three weeks before Christmas, his office won an award for the exhibit from the National Training and Simulation Association. Unable to resist a *Star Trek* pun, he announced that the exhibit was inspiring "the next generation of infrastructure preservationists." Pleasing Dunmire even more, the Carnegie Science Center, in Pittsburgh, expressed interest.

The 2013 government shutdown didn't impair his efforts, either. Given a week off, he filled his time with hobbies that were "very related" to his work. "My life is my job," he explained. "I don't wanna retire. I don't wanna play golf. I don't wanna be on the beach or the mountains. I wanna do my job." He'd been getting the script for LeVar 6 approved, filming parts in Panama, planning to film other parts in Nevada. He'd been looking into another video game about corrosion, and interactive training modules for the Defense Acquisition University. He'd been following progress at Akron's corrosion engineering program—whose first twelve students, a year before they were to graduate, had job offers from California and Texas. He planned to attend their graduation in the spring of 2015. When I asked if he was making any attempt to stay healthy, he laughed it off. When I asked if he was thinking about LeVar 7—even though LeVar had said he was out after six—he said, "Never say never."

In late April 2013, Dunmire and a few of his team went to Oberammergau, Germany, for a three-day corrosion workshop at the NATO School. A few dozen people from allied nations attended. On the first day, after introductions and brief presentations by corrosion officials from the United States, England, France, and Germany, Juergen Czarnecki, the head of the German Ministry of Defense's corrosion program, got up to speak. At the microphone, Czarnecki—the esteemed bearer of a PhD in nuclear physics—went through a long, involved presentation called "The Art of Corrosion." He mentioned reforms to Germany's constitution, reorganization of the Bundeswehr, research at the Wehrwissenschaftliches Institut für Werk- und Betriebsstoffe. Then, at the end of his presentation, he ran through five items on a to-do list. The last was bright red. He said the gathered nations had to use the United States as a model, and called for the adoption of "the Dunmire Process."

There were chuckles in the audience—good chuckles. Dunmire was stunned. He stared at the screen in disbelief—that crazy head-down stare—and began shaking his head. To Czarnecki, he said, "What are you doing?" He was honored, but he thought the phrase wouldn't sell well in Washington. Dunmire couldn't imagine anybody naming anything positive after him, and thought it violated protocol. "You don't name things after career bureaucrats," he said. He also said, "I don't want anything named after me. I want the program to survive."

Since then, though he insists he's no narcissist, the term has grown on him. It's not like he coined it. "It's like the Marshall Plan," he said. "Marshall thought it should be the Truman Plan, and Truman said no, it's the Marshall Plan." Now, at work, he refers to the Dunmire Process regularly. "Oh my God," he'll say while wrestling with a program, "we have to use the Dunmire Process!" America's allies, apparently, feel the same way.

I asked Dunmire to define the Dunmire Process. He said it was imagination mixed with Henri Fayol's management process, and insisted his team—which he called his posse—created it, and deserves the credit. Juergen Czarnecki had meant the Dunmire Process to mean treating corrosion like calculus—incorporating the subject into standard curricula, which Larry Lee had been pushing for years. Larry Lee saw the Dunmire Process as the man in charge riding away on a horse, throwing out off-the-wall ideas, while foot soldiers chased and offered support. Rich Hays, Dunmire's deputy, saw it as more of an art than a science, hinging on nonlinear thinking.

I asked Dunmire, "Is *Star Trek* part of the Dunmire Process?" "Of course *Star Trek* is in there. Of course," he said. "In the original series, what did McCoy say? 'The only constant in the universe is bureaucracy.' It's true, isn't it? He said, 'I don't care if you're a benevolent dictatorship, you have to have bureaucracy to do things.'" Dunmire refined his definition. He said he looks at challenges through a lens formed by his military background, his business background, his work at Heinz, and *Star Trek*. "When James Tiberius Kirk said, 'Risk is our business,' or 'You gotta do the right thing,' or 'Stay in that chair as long as . . .'" He trailed off. Then he restarted. "What would Kirk do?

"Here I am, sixty years old, and something's getting named after me, and I'm still breathing. That to me makes it neat." He giggled. He'd told me before that he never yearned for recognition or awards, only to contribute to this earthly existence such that someday, whether fired or retired, he could look in the mirror and say, "I gave my full measure," rather than "I fucked up." That it wasn't a DOD award didn't matter; that he wasn't the most liked man in the Pentagon was a faraway concern. That his budget remained uncertain rolled right off him. "That's what I find most amusing. I mean, big deal, some guy gets a statue named after him after he's dead. Or a building, or an award. But he's dead. He's dead! But I'm alive. I am alive!"

ACKNOWLEDGMENTS

Because no good adventure ever works out as planned, thanks, first, to Matt Holmes and Jonathon Haradon for going in on that dumb sailboat eight years ago. You two are finer *cabrones* than I.

Enormous thanks to Doron Weber of the Alfred P. Sloan Foundation for supporting works like this that promote the public understanding of science. Humongous thanks, also, to Cindy Scripps and the Scripps Howard Foundation for supporting the Ted Scripps Fellowship in Environmental Journalism at the University of Colorado, where the proposal for this book was born. In the modern media maelstrom, both institutions calmed the seas.

To Lydia Dixon, Kevin Tompsett, Andrew Green, Kevin Davis, Nick Masson, Erin Newton, Danny Inman, Ben Berk, Michael Cody, Rob Gorski, Dan Becker, Matt Nelson, and Jeff Purton: thanks for each having two good ears, two strong legs, one good liver, and a flexible schedule.

To Colin O'Farrell, Ben Miller, Thaddeus Law, Emilie Fetscher, and Mike Beach: thanks for comfortable beds/couches.

To Liz Roberts, Erin Fletcher, Adam Hermans, Walker and Bratton Holmes, Brian Feldman, Cameron Walker, Tasha Eichenseher, Jonathan

Thompson, Tom Yulsman, and Brian Staveley: thank you so much for harsh and generous edits and tiny but crucial fixes.

To Michael Kodas, Jerry Redfern, Leah Goodman, Florence Williams, Dan Baum, Mary Roach, Hillary Rosner, Phil Higgs, Erin Espelie, Evan P. Schneider, Brooke Borel, and, most of all, Mom and Dad: thank you heartily for support and writerly encouragement.

I'm indebted to many fine librarians and archivists, among them Alexia MacClain and Jim Roan of the Smithsonian Institution Libraries; Sue Beates of the Drake Well Museum in Titusville, Pennsylvania; Linda Cheresnowski of the Barbara Morgan Harvey Center for the Study of Oil Heritage at the Charles Suhr Library at Clarion University in Oil City, Pennsylvania; Carol Worster of the Gas Technology Institute; Theo Long of the Biscayne Nature Center; Roger Smith of Florida's Bureau of Archaeological Research; Sarah Martin and Steve Jebson of the UK Met Office; Hilda Kaune of the Institute of Materials, Minerals, and Mining in London, UK; William Davis of the Center for Legislative Archives at the National Archives; and the interlibrary loan staff at University of Colorado.

Thanks to Lynda Sather, Katie Pesznecker, Kate Dugan, and Michelle Egan of the Alyeska Pipeline Service Company; to Cheryl Irwin of the DOD; to Alysa Reich of NACE.

Immense gratitude to Ed Drummond, Stephen Rutherford, David Moffitt, Howard EnDean, Jr., Ralph Nader, Stuart Eynon, and others for scraping through old memories.

Bigger thanks to Alyssha Eve Csük, Bhaskar Neogi, Dan Dunmire, and Ed Laperle for letting me follow you around and ask a lot of dumb questions.

For believing in a new guy and sharpening my proposal, utmost thanks to Richard Morris, at Janklow & Nesbit. And for ironing out my wrinkles and making me look good, huge thanks to Nick Greene, Jofie Ferrari-Adler, and the crew at Simon & Schuster.

ABOUT THE AUTHOR

JONATHAN WALDMAN has written for *Outside, The Washington Post,* and *Mc-Sweeney's,* and also worked as a forklift driver, arborist, summer camp director, sticker salesman, and cook. He grew up in Washington, DC, studied writing at Dartmouth and Boston University's Knight Center for Science Journalism, and was recently a Ted Scripps Fellow in environmental journalism at the University of Colorado. *Rust* is his first book. Visit him at jonnywaldman.com.

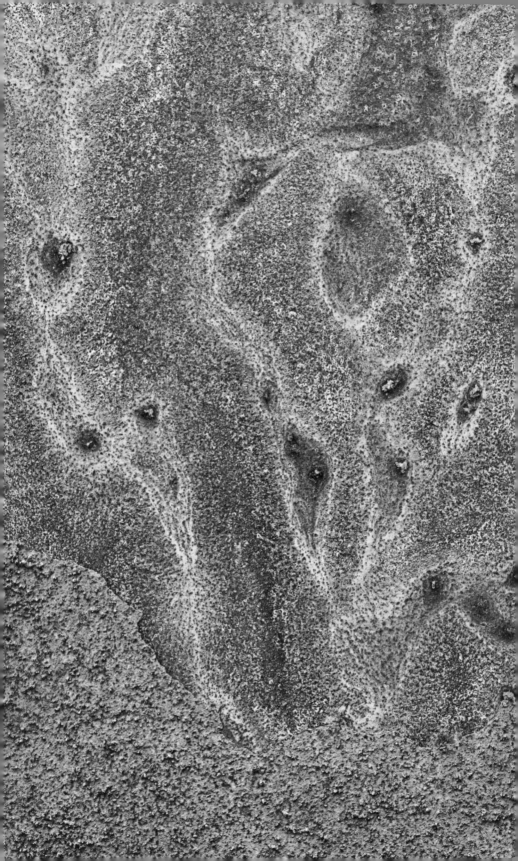